COMPUTER INTEGRATED MANUFACTURING

COMPUTER INTEGRATED MANUFACTURING: Theory and Practice

Daniel T. Koenig
Steinway & Sons
Long Island City, New York

COMPUTER INTEGRATED MANUFACTURING: Theory and Practice

2 3 4 5 6 7 8 9 0 BRBR 9 8 7 6 5 4 3

This book was set in Times Roman by WorldComp. The editors were Carolyn V. Ormes and Michael Schwartz. Cover design by Renée E. Winfield
Printing and binding by Braun-Brumfield, Inc.

A CIP catalog record for this book is available from the British Library.

Library of Congress Cataloging-in-Publication Data

Koenig, Daniel T.
Computer integrated manufacturing : theory and practice / Daniel T. Koenig.
p. cm.
Includes bibliographical references (p.).

1. Computer integrated manufacturing systems. I. Title.
TS166.6.K685 1990
670'.285—dc20 89-28790
ISBN 0-89116-874-5 CIP

As always, to Marilyn, Alan, and Michael for their faith, patience, and encouragement during the preparation of the manuscript.

Contents

Preface

I have been fortunate to be able to observe and partake in the development of computer integrated manufacturing (CIM) since its inception. In those days we called it CAD/CAM. This admission dates me to the late 1960s. Much has happened since then, but the thrust is still the same: to use the computer to augment our ability to optimally manage a manufacturing enterprise. My purpose in writing this book is to define what CIM has evolved into over the years and to show how powerful a tool it can be for competing effectively if applied properly.

This book is not about the intricacies of databases; nor is it about the technical explanations of numerical control. Neither is it about how to program. What it is about is the philosophy of CIM, about how to effectively manage this philosophy so a company can maximize its advantages. To manage effectively, we have to understand the principles of management and have an excellent overview of the topics the manager has to manage. I trust that readers of this book already have experience in managing or at least exposure to the overriding principles of management. This book provides, I hope, the overview of CIM theory that will allow engineers and managers to implement the theory and to use the theory effectively.

I do not claim everything in this book was invented by me. Obviously that would be ridiculous. I am not the originator of the theory. No one individual is. Neither is any one individual responsible for what would be considered a significant segment. Many thousands of people over the past two decades have contributed to the development of CIM, and to them I give my gratitude. This consensus, now reasonably mature in concept, is the basis of the book. The only claim I make, beyond some very minor theory contributions, is an effort to choreograph it and present the CIM philosophy in a reasonable sequence.

Each chapter is complete in itself and describes a particular aspect of CIM. This means the reader does not necessarily need to read the chapter preceding the one of current interest. However, the reader would probably get a little more out of the book by reading it in sequential order. The book is divided into roughly three segments. The first segment is basically an overview of the CIM theory. Why do we do it, and, if we do, what are the real payoffs? The second segment, chapters 5–11, deals with the various modules of CIM. Here I try to explain the theory of components that make up CIM in such detail that the manager will have a sufficient overview of the technology so that the manager can effectively manage its implementation and subsequent operation. Readers should not think that after studying these chapters they will be card-carrying experts in the subject. They will, however, know a whole lot more about it than 99% of the entire business community. I will even go out on a limb and say they will be as knowledgable as the vast majority of computer hardware and systems software personnel.

The last segment of the book is all about managing the CIM philosophy, ranging from what is different, to economic justification, to how to implement CIM in the reader's organization. This segment gets down to the nitty gritty of what to do with CIM. Once the philosophizing is over and done with, how does the manager do something with it? This segment, of course, is based very heavily on my experience.

In more than 20 years of involvement with trying to use computers to make my job as an engineer, and, later, as a technical and then a business manager, easier, I have been fortunate to have had many experiences. Some of these experiences have been very rewarding, and some have not. Many times I have been lucky enough to be in the right place at the right time, where the true action was and still is. I have had opportunities to discuss CIM with some of the finest minds of our time. Some of these people I have even had the honor to call colleagues. It is on this foundation that I drew to write this book. I hope by writing this book I can accurately portray the profound concepts of CIM that promise to revolutionalize how we manufacture products for the benefit of all.

Daniel T. Koenig
Trumbull, Connecticut

Chapter 1

An Overview

Computer-integrated manufacturing is the current buzzword in the world of business. It is viewed as everything from a means of justifying automation to the savior of our standard of living. It may be a little of both or none. What it results in becoming depends on the creativity and knowledge of the users. Computer-integrated manufacturing (CIM) is not a replacement for good management practices. It is a very powerful tool for good managers to use. Like all good tools, the use one makes of it depends on the skill and intellectual capabilities of the user. The purpose of this book is to describe CIM in its totality as I understand it, to impart the basic concepts, and to show its many faceted uses.

The word "integrated" is the most important of the entire lexicon of CIM. It means coming together, the antithesis of stand alone. It is the cornerstone of the CIM philosophy. It implies that all functions, activities, decisions, and questions be acted on not solely for the immediate task at hand but for what it means for the entire entity. This is truly the axiom, "The whole equals the sum of the parts." CIM means we use computers to assist in making decisions in a manner that takes into account, first, what that choice means to the entire business and, second, what it means to the specific subset. This requires integration, which means each person in the organization has to think in global terms and not strictly within their sphere of expertise. When this occurs, we have synergism and optimal results. It is a fact that the more optimal results an organization achieves, the more likely it will be a winner.

What integration requires is a successful communication channel, one that is accurate, fast, and current. This leads to the hallmark of computer use in CIM theory:

the common database, the sharing of the same information for different purposes by diverse subsets of the organization. The theory then requires easy access to data by all for whatever reason the user perceives. This is what makes CIM different than ordinary uses of computers. We traverse from singular-use information systems to multi-use, while, at the same time, the multi-use is as efficient to the specific user as the traditional stand-alone systems.

This book describes the many uses of computers in business entities. In many cases, the use may seem similar to common stand-alone systems. Indeed, they may be identical. However, the overriding concept is to show how they interrelate with each other and why integration is a better solution. Many different topics are presented, and CIM is viewed from many different directions.

Throughout, the purpose is to show the need for integrated subparts for implementation of computers within a business to be optimum.

The common database is essential in understanding the theory of CIM. What exactly is in a common database? There is no exact answer. The content is very dependent on the nature of the business or activity in which the organization is involved. However, for manufacturing concerns, we can outline what is contained in the common database through the CAD/CAM triad. The CAD/CAM triad describes the integrated activities that are either computer controlled or assisted in making a company's products or developing its service offerings. Triad implies three; hence, there are three equally important phases in a computer-dominated manufacturing control structure:

1. machine/process control
2. design and planning control
3. production and measurement control

All concepts of management control and decision making are represented in the CAD/CAM triad. For example, contained within the machine/process control triad are all the strategies and design of machine tool or process numerical control. Included would be, for instance, N/C, CNC, FMS, robotics, or family of parts programming. The design and planning control triad deals with computer systems developed for data collection systems, process planning, automated design and drafting systems, methods and time standards systems, and other similar up-front activities. The third triad, production and measurement control, is dominated by manufacturing resources planning (MRP II), just in time (JIT), material handling systems, and a large variety of business information and financial systems. All phases of the CAD/CAM triad are equally dependent on the others and are also mutually supportive. This will become obvious in later sections of this text. Since the legs of the CAD/CAM triad are both supportive and dependent, it becomes obvious that shared information not only is desired, it is necessary for optimum results, and hence, the common database requirement of CIM.

With common databases it becomes apparent that easy quick access to the data is required for it to be a useful tool. All practitioners of management agree that information routed to the proper person at the proper time is key for success. So, having data in a common database environment is not sufficient in itself. The data must be available and must be easily and quickly found by any user. In CIM systems, this is accomplished through group technology classification and coding systems. Many refer to group technology as the DNA/RNA of CIM. It is a method of unlocking the potential of CIM such that it truly allows all members of a company's organization to expeditiously search a database for information they need to do their jobs. Group technology also offers the potential to use developments in artificial intelligence for optimizing manufacturing planning. The use of group technology in a CIM environment is an important element in understanding the value of CIM. It is covered extensively in subsequent sections of this text.

CIM is often considered a synonym for the so-called "factory of the future." This is not entirely correct, but, certainly, future aspects are important. If the factory of the future implies the most efficient and productive facility possible, then CIM is implied. But it does not have to lie somewhere in the future. It can be here now, with intelligent application of CIM principles. Factory of the future means computer-driven processes and information systems. It means managers can have current accurate information in a dynamic mode, thus allowing decisions to be made with real-time information, giving the company the best possible chance for success. Note the emphasis on information. The optimum factory, or organization, really thrives on communication excellence. This is the domain of CIM and is how we achieve winning situations. The range of technical and managerial systems, processes, and techniques is the content of this book. Understanding these subsets and then seeing how they are interrelated and linked into a synergistic whole is the purpose of this text.

Chapter 2

The Promise of CIM

Let us now look at what CIM can be: the promise that it offers to business success and the reasons why the interest in this technology is so intense.

To understand the promise, we must investigate what CIM is for, what it encompasses, and what it means for a business. First, some terminology:

- CIM: Computer-integrated manufacturing (circa 1982)
- CAD/CAM: Computer-aided design/computer aided manufacturing (circa 1972)

They really mean the same thing. But, most uninformed people thought the slash meant a barrier between manufacturing and engineering and, hence, the need for a new definition. It is entirely correct to use the terms interchangeably. Both terms imply that the computer becomes intimately involved in the workings of the major components of a business. Also, since the computer is involved, we correctly imply that the goal is improved effectiveness of the enterprise. In fact, we must always keep in mind that the only reason for CIM is to improve profitability of the enterprise. All other reasons for CIM are secondary. All other reasons given for achieving a CIM solution are subsets for achieving improved profitability.

CIM improves profitability by integrating all functions to minimize costs and by preventing suboptimization of functions. In essence, CIM is the optimization of communication excellence. CIM optimizes communication because it exists in a dynamic world, due to the computer, and not in the static world we were relegated to only 10–15 years ago. We are not really radicalizing our approach to business control.

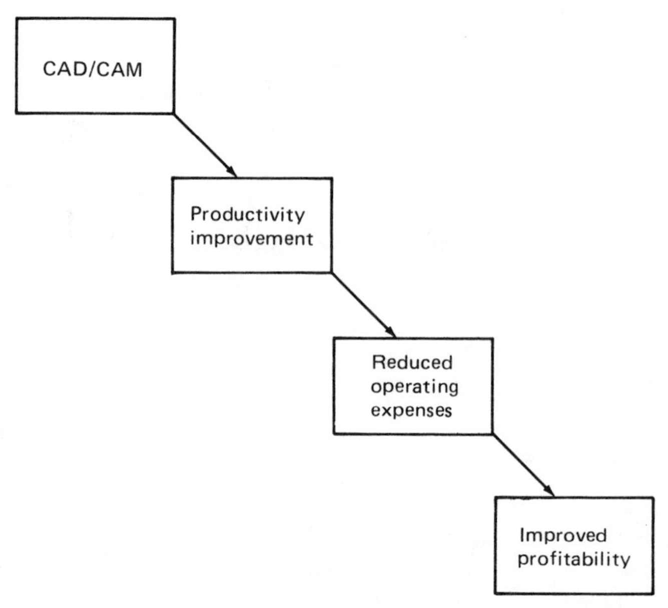

Figure 2.1 Benefits of CAD/CAM. (From Daniel T. Koenig, *Manufacturing Engineering: Principles for Optimization,* Hemisphere, Washington, D.C., 1987.)

We are only doing what we would have wished to do but could not do before. The approach to what we would like to accomplish, in a dynamic mode, is as old as the Industrial Revolution itself. It is called the "Manufacturing System."

The Manufacturing System has been optimized as best as it could in the static mode for generations. CIM allows for dynamic optimization. Let us look at the Manufacturing System. It consists of seven integrated steps (see Table 2.1). All companies offering products or services follow these steps either consciously or unconsciously.

The ultimate goal for CIM is the linkage of all seven steps in an automated integrated fashion. For example, a firm would benefit from a singular database for step 1 (obtain product specification) to step 2 (design a method for producing the product),

Table 2.1 The Task of CIM Is To Automate and Integrate the Manufacturing System

Step	Action
1.	Obtain product specification.
2.	Design a method for producing the product, including design and purchase of equipments/processes to produce, if required.
3.	Schedule to produce.
4.	Purchase raw materials in accordance with the schedule.
5.	Produce in the factory.
6.	Monitor results for technical compliance and cost control.
7.	Ship the completed product to the customer.

thus greatly reducing the probability of information transmission error between the designer and the manufacturing engineer. It is important that the seven steps be thoroughly understood. Let us define them and show what we mean by existing in a dynamic mode instead of the usual static mode. The techniques of obtaining a dynamic mode will be defined in later sections of this text.

Step 1. Obtain Product Specification. This is the design step. It is either done by the firm, or it is supplied by the customer. Sometimes it is a combination of both. In this step the particulars as to what will be made are thoroughly defined. Here, the content of the product is defined in a manner that can be understood by the firm so that the required manufacturing functions can be performed. By "design step" we mean total design. Materials and geometry are created, and all analyses related to assuring that the product will do exactly what it is supposed to are done.

"Static mode" means applying engineering theory to the concept of design, that is, performing the calculations, creating documentation (usually in the form of drawings and instructions), and submitting these results to some form of retrievable archival system. Sometimes, computers are used to aid in the calculations and drawings. There is no formal linkage with the recipients of this information.

In the dynamic mode there is a linkage with the recipients of this information. The creators of what the firm is to produce are directly linked to the functions that will manage the production of the product. In fact, it is sometimes difficult to differentiate between the creators of the design and the producers of the actual artifact. In the dynamic mode we are acting in unison, not the old style, stand alone.

Step 2. Design a Method for Producing the Product. This step is considered the classic manufacturing engineering activity step of the Manufacturing System. Here, engineers define what they have received from design engineering into plans, methods, and procedures for manufacturing. They also break down the material needs by sequence of use and quantities required. Many times decisions have to be made as to whether or not present equipment is adequate to produce the desired product, hence the need to design and procure new machine tools and equipment. As in step 1, the work of this step is stored in accessible archives for later use. Many times, computers are used for the analyses required and to store and classify data.

In the static mode, manufacturing engineers often act without consulting design engineers. They make the assumption (often dangerous) that all the information they require has been adequately defined for them by their design brethren. They create their own archives and their own interpretations of product functionality, as required. This, as we know, can and all too often creates product performance problems and, hence, profitability problems.

If we had the desired dynamic mode, then there would be continuous communication between these two functions (actually between all functions). In fact, much of the work traditionally performed by manufacturing engineers would be performed by design engineers. For example, manufacturing engineers usually do the programming of the numerical control machine tools. This typically means transcribing the geometry of parts to be made into tool path algorithms. This is really redundant. The design engineers had to define the geometries in the first place. Why could they not define geometries in a manner that could be interpreted by the machine tool without the intervention of the manufacturing engineer? This is possible and can be done in a dynamic mode. The two engineers can become one.

In a dynamic mode the two archives can become one, and we have then eliminated many sources of errors. Communication between functions reaches the point of perfection because the multitudes of similar data sources are replaced with many fewer coordinated ones.

Step 3. Schedule to Produce. This is the tactical step of preparing long-range and short-range loadings of the work stations throughout the company, its vendors, and, perhaps, the customer (typically, for liaison information, e.g., foundation and electrical interfaces for a complex machine structure of which the company is only producing a subset of the entirety). Here, engineers use the capacity calculations, cycle times, and process requirements provided by various manufacturing engineering functions to develop schedules for the factory.

In the static mode we see "war rooms" with bar charts and such decorating the walls. There are clerks making changes to schedules and determining how those changes affect other schedules. The paper flow is enormous, and much of it is in a shorthand language known only to the schedulers themselves. We have literally platoons of checkers whose job it is to tour the factory to test schedule compliance and accuracy so schedule revisions can be prepared. There is virtually no contact with other parts of the business who had originated the data, no questioning of the accuracy of the capacities and cycle time, no thoughts as to whether or not the design is being faithfully brought to life. Occasionally, computers are used to store data, but independently of data from any other source, and to do arithmetic operations pertaining to schedules.

In the desired dynamic mode, there are no massive bar charts on the wall. Instead, the schedulers use computer terminals to access the data from manufacturing engineering, add customer due dates, check the design engineering database for peculiarities, and issue schedules for the various work stations. Not only do they issue schedules, but the schedules are interrelated such that the reservation system does not have more than one part being worked on at one time at one work station. We also have no need for the platoon of checkers because a dynamic feedback system is in existence. In essence, the scheduling activity is tied into an accurate communication system that encompasses the entire company.

Step 4. Purchase Raw Materials in Accordance With the Schedule. Here we find the core of the traditional materials function. In this step, purchasing agents procure raw materials at the best possible price and at the required quality levels for delivery in accordance with factory needs. By raw materials we mean any component needed to produce the company's products. This means that the material can be anything from completed subassemblies to ore just out of the mine.

In the static mode, purchasing agents work with due-date schedules, drawings, quality requirements, and, occasionally the assistance of technical personnel. Their work consists mostly of finding the best financial terms and only sometimes with the intimacy of when the material really needs to be delivered to the factory. Since schedules often vary, their priority is to ensure stock is on hand when shop operations are ready to use it. So, the policy of reserve stock becomes an important factor. Thus, the just-in-case mode prevails.

In a dynamic mode, purchasing is tied directly to the schedule. This becomes the core of MRP (materials requirements planning, sometimes known as little MRP). Ordering is done by the schedule, and purchasing's role is that of coordinator and establisher of longer-term contractual agreements with vendors. Purchasing agents directly access the design databases to monitor changes. They then electronically make changes to vendor information pertaining to in-process buys. This not only speeds dealings with vendors but also ensures that the material received is coincident with current needs. A similar linkage with the scheduling database also ensures that material only arrives when actually needed. This is conducive to the establishment of the so-called just-in-time philosophy.

Step 5. Produce in the Factory. Here we find the traditional shop operations function. Material is dispatched, work stations designated, and labor is applied to produce the product. This is the domain of the supervisor and the hourly paid work force.

In the static mode the supervisor assigns work from daily schedules received from production control. Usually at the start of the day it is correct. By that is meant it is updated to reflect the actual progress of the previous day. It is hoped that variations in performing yesterday's activities are reflected in today's. Often, the supervisor finds a need to make modifications right from the start because of machine malfunction, for instance, or absenteeism or material discrepancies. The supervisor usually has good intentions of informing production control of these adjustments but may not do it in a timely fashion. In practice, for all of these reasons, schedules rarely match the actual. To overcome these variations, some say ordained from heaven, shop operations tends to lobby for extended cycle times to compensate. This always leads to additional material and labor costs, plus the need to employ hot lists to get deliveries accomplished on time.

In the better environment of the dynamic mode, scheduling is real time. Scheduling of individual work stations is done by the computer by using interactive terminals. These terminals query current status and only dispatch the next job when the previous one is complete. The schedule then is constantly being modified to be in concert with reality. Shop operations then has current real-time information to match against pre-set requirements. Actions are taken for correction when they are really needed and not after the fact. This significantly reduces cycle time. The database also feeds information to the purchasing agent's database and coordinates deliveries of materials when they are actually needed. Here again we see the benefits of linked systems providing accurate information.

In the dynamic mode the hourly worker is tied directly into the decision-making loop. This interactive capability tends to solicit better ways of doing the job all across the board. In the dynamic mode the supervisor is freed of the less productive task of being a status verifier to a manager who can analyze the current situation for strategies leading to improvements. The dynamic shop operations mode allows us to use more effectively the creative talents of all employees.

Step 6. Monitor Results for Technical Compliance and Cost Control. This is the quality control and financial cost accounting phase of producing the product. Here we are comparing actual performance to what the company had planned to do. Also, information gained is applied to corrective action plans. While the finance and manufacturing departments have different goals for the information, they are dependent on the same data. The quality control department measures discrepancies for an understanding as to what it means for product functionality. At the same time, the cost accounting department is measuring the same discrepancies for their effects on product cost.

In the static mode, quality control and cost accounting people go about their business independently of each other. Quality control sets up inspection plans based on what are deemed critical operations and gathers data from them. They also gather data on a random basis from reported discrepancies as they occur. Sometimes the discrepancy reports are forwarded to cost accounting people after corrections are made so that costs can be calculated and added into operating expenses. Usually, though, cost accounting does not rely on quality control for cost data. They traditionally collect time sheets for labor performed, calculate labor costs by rates per hour, and compare actual labor expended to planned labor from the manufacturing engineering people. Material costs are added in on the basis of withdrawals from the stock rooms. Some of this activity is done by a computer, usually the arithmetic activities. Typically, there is no crossover of databases. In fact, data collection is done independently, the quality control usually from tally sheets at the work stations and cost data primarily are derived from the payroll time sheets. In a static environment, these functions exist in separate universes hardly ever even realizing that they have common needs.

The world of the dynamic mode is one of unification, one where conscious efforts are made to find commonality of needs even though those needs support different requirements. With modern computer technology, we find that data collection systems can be employed to collect data only once and only in one format that serves the needs of many users. This is an obvious productivity saver for those required to provide the information (most often the harried personnel of shop operations). This not only saves time, but, because it is easier to do, it enhances the quality and accuracy of the information. Another significant improvement only available in a dynamic environment is the real-time factor. The traditional way of monitoring results for quality control is by a daily count for simple things and weekly for more complex, with a weekly summary for the total. This meant that management could not react effectively when problems were actually occurring. Cost accounting was even further encumbered. Since payroll is a weekly activity at best, it was hardly possible to collect and analyze costs for less than 2-week intervals. Usually, a monthly interval was selected because of the work involved in sorting the data. With data collection, these constraints are eliminated. Data are collected in real time and available to users virtually instantaneously. It is now possible to have a blow-by-blow description of what is going on during the manufacturing process.

The dynamic mode implies shared databases. This is entirely correct. We have data being input by various functions and available to all. Now the entire set of functions required to run a business are capable of working as an integrated team instead of as independent and sometimes conflicting teams.

Step 7. Ship the Completed Product to the Customer. This is the materials function again. Here, the company receives the completed product, packages it, and dispatches it to the customer. Activities occurring center about arranging for transportation and updating the master schedule to show the complete job. The shipping information is also given to the finance department so customers can be billed.

In the static environment, the shipping department works with tally sheets and a shipping plan conceived by the marketing department. The tally sheets portend to state accurately what they will receive from final assembly to ship during the current shipping period. The marketing plan states what the master schedule predicted or desired to have shipped during the current period. They hardly ever match. There is little if any feedback to update the master schedule, which is the driver of the shipping plan. This means there are always efforts to find other products close enough to completion to be expedited so that sales volumes can be maintained. Without the sales volumes at satisfactory levels, cash flow cannot be maintained; hence, the profitability is in danger. Other problems manifest themselves owing to inaccurate information. Particularly irksome is the "sawtooth curve" of shipping activity. Since the beginning of the shipping period is dominated by a search for product to be shipped, very little is shipped. The work force is waiting for heroic efforts to be accomplished by the assemblers to expedite product completions. The assemblers, in turn, are at the mercy of other operations down the line, and so on. Finally, goods start to arrive, and it is now the shipping department's turn to do heroic deeds. Hence, we see the "sawtooth" load develop. At the end of the shipping period, with luck and hard work, the sales volume shipped is accomplished, and everyone congratulates each other. But, at what cost: high overtime costs, further muddling of the schedule, less than satisfied customers, and worn-out employees who have to rest before the next battle begins.

In the dynamic mode, none of this has to happen. With real-time data collection, all schedules, including the marketing schedule, can be kept current with reality. The surprise element can be minimized to the point of nonexistence. Shipping would not be scurrying around for products to ship. That would be known through adjustments to schedule beforehand. In fact, through timely receipt of information, corrective actions could be taken in a more timely manner, and there would not be as many deviations as before. With accurate information through real-time data collection, the marketing department could inform customers of delivery problems much earlier so the customers can take actions to mitigate the effects. Information commonly available to all certainly makes the company more effective and, hence, profitable.

We can see that the seven steps of the Manufacturing System require a linkage to be performed optimally. The dynamic mode made possible by a CIM solution offers that capability. We can go a step further and state that CIM, then, is the means by which we link the seven steps in the optimum manner. This is another way of saying we are achieving "communication excellence," and this will produce our goal of superior profitability.

The ultimate goal for CIM has to be the linkage of all seven steps in an integrated automated fashion. The business system, from order entry to shipping of product to customers, has to be linked by common information networks. This requires the three legs of the CAD/CAM triad to truly be linked. While there may be on the surface no apparent linkage between the programming of an N/C machine and cost accounting, there really is, and we must understand that so as to not suboptimize one at the expense of the other.

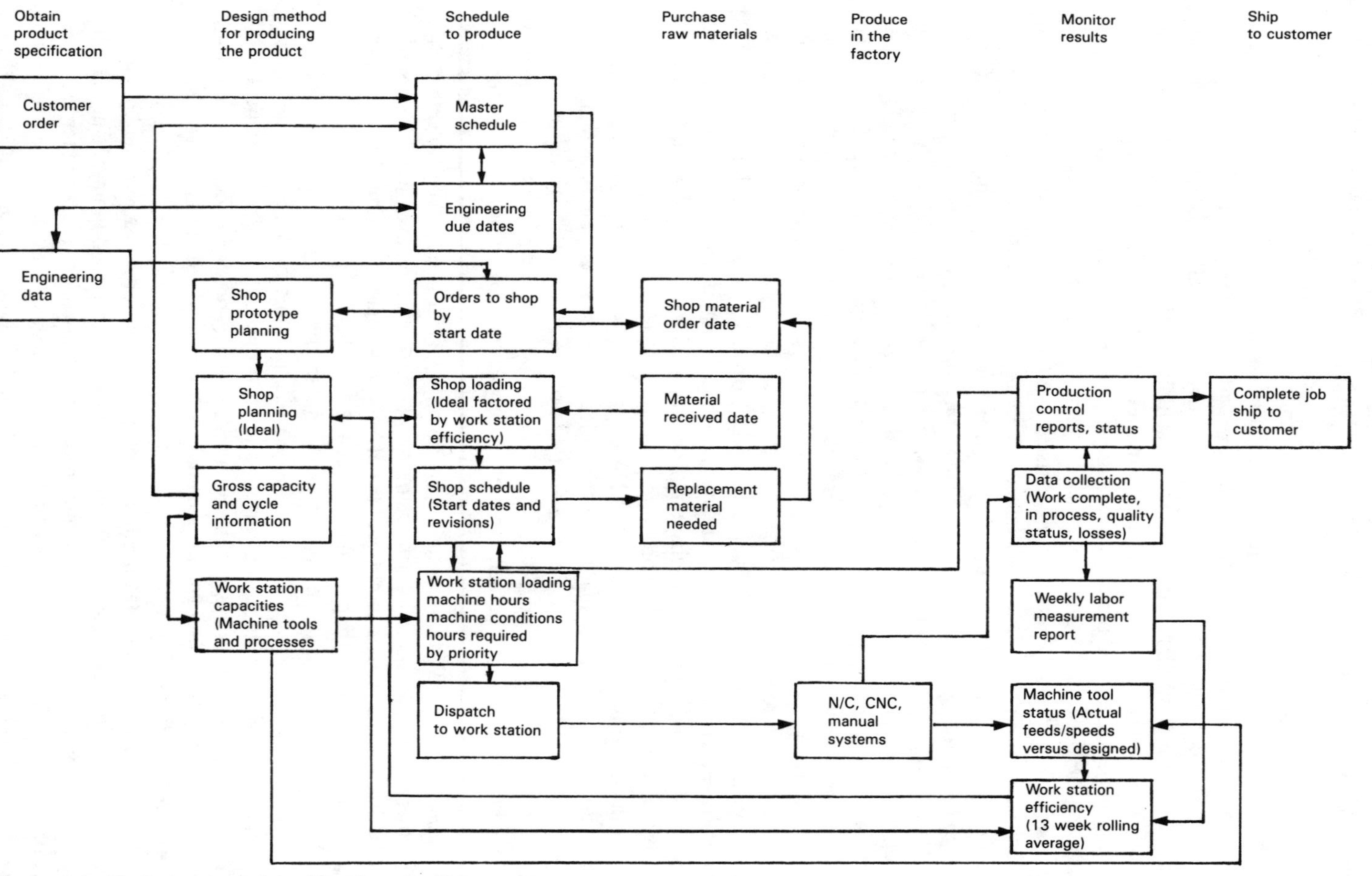

Figure 2.2 Typical variant of the Manufacturing System.

If linkages are going to occur, then all functions of the organization must work with the same information and the same definitions. Working with the same information does not mean that individual components' sub-purposes are the same, but the ultimate goal would be similar. For example, the finance department may not need to know the extent of finished part tolerance drift deemed acceptable by the manufacturing engineering department. All they would need to know is whether the parts are usable so labor and material costs could be allocated to the proper cost categories. The ultimate goal both are interested in is the yield factor of good parts. This comes from common, or shared, information. In computerese, this is referred to as a common database, which will be discussed in later sections. For now it is sufficient to say that common databases are structured so that all users find the information formatted (meaning the way information is filed and presented) in a way they can understand and use. It is also a requirement that the information be input in a manner that is considerate of the use that it may be put to by others. An illustrative example may be useful here.

Let us look at an example of doing things differently because we want to use the same information for engineering and manufacturing needs. We want to optimize linkages between these business functions. We want to create a database for all and not just an output of one function that is oblivious to the uses to be made of it by others. We want synergism. Figure 2.3 shows in a very simplified manner the old way and the new integrated, linked, way of providing information.

On the left we see a geometric representation of a dowel as it normally would be described by an engineering drawing. This was the state-of-the-art in non-CIM days. The dimensions are perfectly clear, and it would be possible for the manufacturing department to produce (providing material was also specified). The right side of the figure shows the CIM way of defining the same dowel. What is different? The designer is thinking about what other users may want to do with the information, which, in this case is to manufacture the dowel. The designer has structured the information with respect to datum points. It is still the same seven inches long by two inches diameter dowel, but now the designer is considering how a machine, or coordinate measuring device, or perhaps a robot, may visualize the part. The designer is thinking how others may need to use the information. Here we see the need for the designer to think beyond his or her own needs for describing the part. The need is recognized to perceive a standard so others can easily use the information. This leads to significant synergism: a very basic way of improving costs. No longer does the engineering department finish a drawing and simply pass it on to the manufacturing department without any concern or thought as to how the manufacturing department will use the information. In this example, if the manufacturing department wanted to use the engineering drawing, it would first have to create a datum point so that raw material stocks could be placed on a lathe. This is redundant work. The engineering department had to do a similar thing to lay out the part. Again, we see benefits immediately. Why do something twice? Indeed, with common databases and functions linked, there is no need.

Linkages between functions create opportunities to improve productivity and, hence, profitability. Productivity has to happen by CIM techniques, otherwise it is a useless exercise. We have to ask continuously, "How does CIM improve profitability?" If we do not know the answer, then we have to find it before we proceed. I believe the

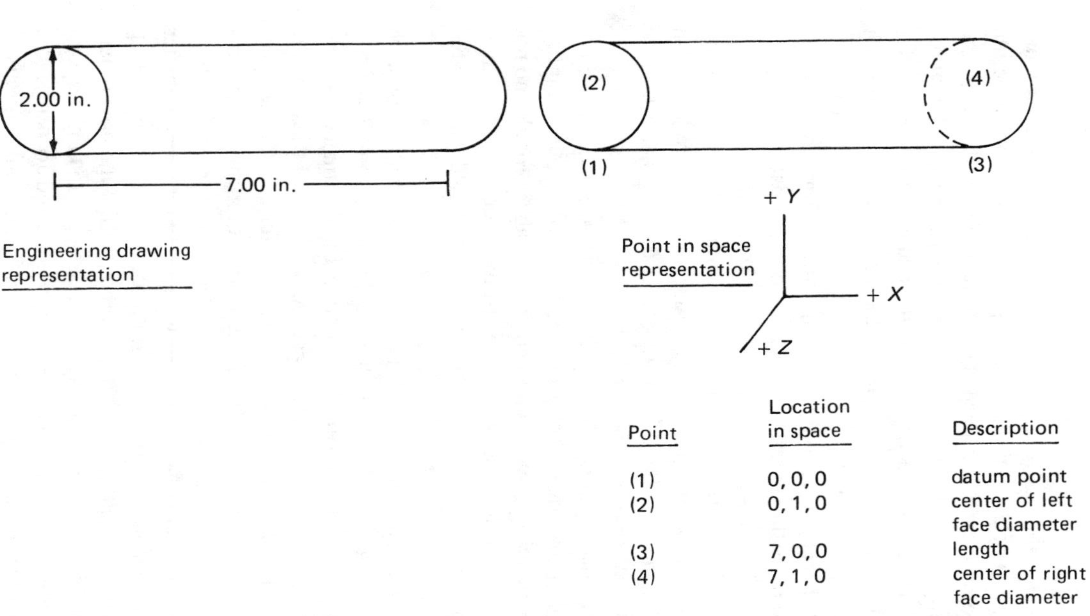

Point	Location in space	Description
(1)	0, 0, 0	datum point
(2)	0, 1, 0	center of left face diameter
(3)	7, 0, 0	length
(4)	7, 1, 0	center of right face diameter

Figure 2.3 Converting from engineering drawing to database. (From Daniel T. Koenig, *Manufacturing Engineering: Principles for Optimization,* Hemisphere, Washington, D.C., 1987.)

answer lies in the Manufacturing System. We see that in each of the seven steps it is information we seek. Therefore, the common denominator is information giving and getting. So, information is the linkage between the seven steps. Our previous examples certainly support that contention.

Information requirements for all seven steps are essentially the same:

- What is it?
- How is it supposed to work?
- Who has to do the work?
- When does the work have to be done?
- Why does it have to be done in a certain way?

Each of the organization's functions asks these questions pertaining to what it does, but its information source is common to all. The finance department has to ask these questions so that proper accounts payable and receivables along with cost tracking can be established. We can see that even the employee relations department needs can be supported by a common information source. The staffing of a component by necessity is driven by the type of work being contemplated.

If all components have equal access to information needs to effectively run the organization, then it is more likely that a unified and coordinated result will occur. This has always been the goal. But, since information gathering and disseminating has been so chaotic, or unorganized in the past, we have not been close to optimal. With the computer and a database equally accessible to all, optimization is distinctly possible.

This is the essence of CIM: "excellence in communication."

Excellence in communication is achieved through common databases. This is the difference between "islands of automation" and CIM. Figure 2.4 symbolizes how

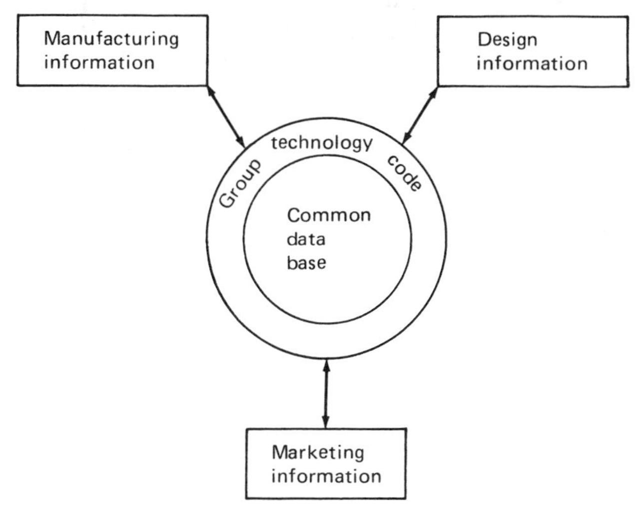

Figure 2.4 Inputs and outputs of the group technology coded common database. (From Daniel T. Koenig, *Manufacturing Engineering: Principles for Optimization,* Hemisphere, Washington, D.C., 1987.)

productivity occurs through sharing of information by common databases. Here we see the major information paths available to a manufacturer: design information, manufacturing information, and marketing information. This set of information is stored in a common database accessible to all. Note, however, the common database is guarded by something called "group technology code." This is both symbolic and rational. It is symbolic in that it represents a methodology of access to the database that protects from unauthorized changes and deletions. It is rational in that with so much information stored, there has to be a logical manner of entry and comparison of information for the system to work, sort of a catalog system. We will describe group technology coding extensively in a later section of this text. Figure 2.4 shows that information is sent and delivered from all information sources, and it is shared information based on needs of the user.

The promise of CIM is that of significantly improved profitability. So far, the description of how that occurs has been focused, and rightly so, on communication. Now we will change direction slightly to put in perspective the basic area where that information is used and how it, too, is linked in an interdependent manner.

To control the entirety of a manufacturing or services-providing company, we have to know the basic information processes required. This is defined by the CAD/CAM triad, which was briefly introduced in chapter 1.

The triad is like an equilateral triangle (see Figure 2.5). All three legs have to be there for the triangle to exist. The same is true for CIM. All three branches of the CAD/CAM triad have to exist for CIM to exist. These branches are machine/process control, design and planning control, and production and measurements control.

All companies that produce goods and services, as described previously, do indeed comply with the seven steps of the Manufacturing System. The CAD/CAM triad is another way of expressing the seven steps, only this time it is described by classification of activity related to planning (design and planning control), doing (machine/process control), and measuring results (production and measurements control). This is done

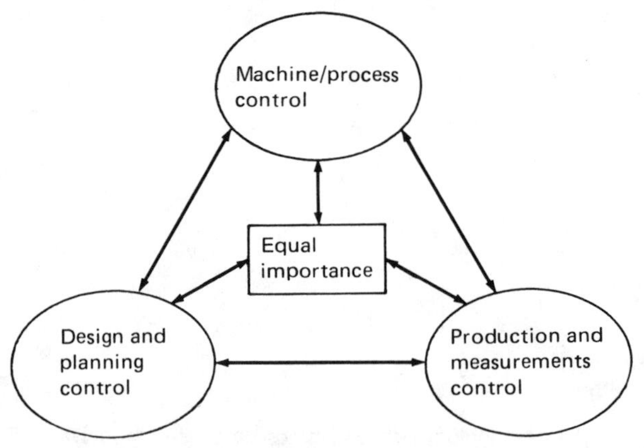

Figure 2.5 The CAD/CAM Triad. (From Daniel T. Koenig, *Manufacturing Engineering: Principles for Optimization,* Hemisphere, Washington, D.C., 1987.)

so it is easier to evaluate hardware and software impacts on overall CIM implementation strategies.

Following the equilateral triangle analogy, we can see that to maintain the triangle as an equilateral and to increase the perimeter of the triangle, all sides have to grow equally. Likewise, for CIM to grow within an organization, care has to be exercised to be sure that the needs for all aspects grow and mature together. This means we cannot have a major emphasis on numerical control machines at the expense of planning systems and purchasing systems. Things have to be kept in balance. Implementation of CIM requires the linking of the seven steps. By visualizing the seven steps in the CAD/CAM triad, it is easier to keep implementation in balance and also to teach the interrelationships existing in CIM.

Visualizing interrelationships is very important if integration is to occur. Too many in responsible positions are fascinated by what I call the "whiz bangs" of CIM (e.g., the N/C, CNC, or robots), devices often associated with advanced computer applications in manufacturing. These are important for productivity improvement. However, they are only the tip of the iceberg, and, if management concentrates only on these, then they will miss most of the benefits of CIM. In fact, they will not achieve CIM at all but rather a series of "islands of automation." We have to wean people from this erroneous interpretation of CIM. The factory of the future, the factory of optimum productivity, is one where automated equipment may and probably will be employed; but, primarily, it is one where excellent, complete, instantaneous communication is the standard and decisions are made on a real-time basis. The CAD/CAM triad is very useful in instructing people as to the integrated necessity for successful applications of computers in business.

Implementing N/C or CNC without consideration of planning or measuring systems only makes use of the machine/process control triad. Certainly, this is not optimum for the following reason: We would only be affecting from 5% to 20% of the cycle time factor of producing the product. The goal has to be to affect the entire cycle time factor. In job shops, which make up 70% of all manufacturing, only 5% of the portal-to-portal time to make a product is value added (20% for flow manufacturing). By "value added" is meant the physical or chemical conversion of raw material to finished parts and the assembly of those parts into a finished product. Even if we totally optimize the value-added portion of production, we have not accomplished much. What of the other 95%? It is so apparent that attention must also be paid to the other two parts of the triad if true improvement of significant magnitude is to occur. Planning and measurement have to be computerized as much as the machines and processes.

For CIM to achieve the promise of significant profitability improvement, all three legs of the CAD/CAM triad must be employed, and they have to be in synchronization. While machine/process control is the most visible and the catalyst that sets the need for the other two, all three must be in balance. Sharing information from the same databases, the triad acts in unison to create the true CIM world. The promise of CIM states that we can expand the three sides of the equilateral triangle called the CAD/CAM triad, and, by doing so, we can achieve a dynamic real-time solution to the total business optimization problem.

Chapter 3

The Databases: Architecture Types and Control Concepts

The database is the most abused concept of CIM. The term itself is used to convey a sense of mystery and awe surrounding computer utilization. We hear things such as disk storage, on-line access, data dependence, input/output (I/O) queue, and response times with respect to databases. These are all necessary and may even be important in certain situations. However, this is all tactical and not of strategic significance. Let us state what a database is. A database is an organized assembly of information that a user or group of users can access for whatever purpose they have. That is it! Any other function claimed for databases is hype and irrelevant.

It is the strategic concept of databases that managers have to be concerned with. We know that the prime purpose of CIM is to improve the profit potential through communication excellence by means of integration of the Manufacturing System. The database is the vehicle of storing, sorting, and recalling information for the entire system. The strategic question then becomes how the database should be structured. To answer this question, we must first understand the uses of the particular information system; that is, what we, the users, want to do with it. Then we must understand how databases may be organized external to the computer and, to a certain extent, internal to the computer. Knowing these three things, managers can then make decisions pertaining to the types of database external and internal architectures to select.

When we go about the task of selecting a database system to support a particular CIM strategy, the following criteria have to be evaluated. This constitutes understanding the uses the database must support.

• *Type of data to be stored.* This is primarily an evaluation of whether data are numerical, descriptive, or some combination of both. Numerical data tend to require less disk storage and computer memory space than prose. Understanding type of data stored is important for sizing and performance (response times) evaluations.

• *Amount of data to be stored.* This is a loaded question. Rarely will users fully understand the amount of data they will generate when their system is actually operational. But this does not relieve the responsibility of management to estimate the size of the system. A practical rule of thumb would be to make the estimate and then double or triple it.

• *Number of data files required.* This is another parameter for sizing the system. The number of files can be loosely paralleled with the number of activities that must occur to conduct the specific business function. For example, when doing order entry, how many manual files are entered with the information and how many actions are required to fully ledger the order? Each one of these actions could conceivably be a data file in the database. The advice is to count every manual transaction, no matter how trivial, as an independent entry. This is the conservative approach and will result in large enough estimated file quantities in the sizing exercise.

• *Who will use the data?* This determines the number of access stations necessary. We call these terminals or ports. The number of terminals is an important factor in sizing the computer hardware requirements. Another consideration would be the degree of computer literacy to be expected. The less literate the users are, the more tutorial the programs employed will have to be. This tends to increase the size of the program and, hence, the size of the database.

• *Access speed required.* Once the number of terminals is determined, the next logical step is to understand who the users are. This will define how quickly the system should respond to user inputs. Speed of response depends on the user's need. For example, a design engineer probably would be satisfied with a minute to several minutes response time for a complex activity such as finite element stress analysis of a structural component. At the same time, an operator at a lathe would want response in seconds for inputs into a statistical process control program telling him or her if the component is being produced within design tolerances.

• *Flexibility of change.* This is a very important consideration for what type of internal architecture will be selected. Managers must be realistic about the type of activities they do. Are they repetitive and routine (i.e., predictable), or are they indeterminant? Usually, business operations are a combination of both. However, they tend to favor one extreme or the other. The task is to determine which one predominates. The internal architecture must support the predominant need.

• *Access control.* This tends to be an administrative manner. But it is important to consider who will be able to access the data and to what extent. This is a way of protecting data and maintaining its legitimacy (i.e., believability) over a period of time.

• *How maintained?* Again, this is primarily an administrative manner. However, if the needs of the organization require a very interactive tutorial type of CIM system, then the database will be more complex and probably will require a computer systems engineer on staff to maintain the system.

Let us reiterate once more: Databases are storage and retrieval mechanisms, nothing more. Think of them as a library of information stored electronically instead of by books, periodicals, letters, and other hard-copy methods. The problem with hard-copy methods is that they are difficult to manipulate. Recall from your own experiences the tedium and extended time it took to do research for class papers in a library that even had a good card reference system. Also think how difficult it is to keep all that data current as changes occur. This is the reason electronic databases came into being. The computer offers a very fast and practical way of storing information and updating it. It simply means accessing the information on a terminal and making the required change. This can be done from multiple, even remote, locations without the bother of physically retrieving the hard-copy record. Also, at the time of the change, all associated records and affected records can be simultaneously updated. Try doing that with a manual system! The electronic database offers advantages, such as those mentioned, that our librarians and archivists could not even dream of a decade or two ago. Electronic databases allow us to enter what many call the "information age."

Once we know what the criteria of the database are in a specific situation, we can select the internal and external architecture to best match the conditions. All computers have internal architectures that instruct them how to handle data. We do not mean the core internal structures and instructions as to how the computer sends electronic signals into and out of its central processing unit (CPU). For this discussion we mean how the computer manipulates data, how it is instructed to find data, and how the data are used. These are master programs about which applications programs can be constructed. At present, there are three common internal architectures that are used to store data and set the rules for applications programs (software): hierarchical, network, and relational. These will be described shortly. There are many others beyond these three, but these are the basics. Undoubtedly, architectures will continue to be developed, and these three will become obsolete; however, they serve our demonstrative purposes well.

The external architectures refer to how the computer system is structured to serve the user's needs. This is the linkage to the CIM strategy. Here we are interested in how data are assembled and how interactive they are with individual or group users. External architecture relates to the information flow of the business entity. If common data are desirable, as is implied for a CIM system, then the organization of that data is critically important. In a later section of this chapter we will discuss how that is accomplished through external architecture strategies. For right now it is sufficient to know that there are two types of architectures, and they come together through the distribution of internal information external of the computer hardware schema.

At present, there are three types of databases in general use to manipulate applications programs within a computer. We should clarify that we mean in "general" use. These are the databases that structure the common data storage and are acted on by the user by applications programs. Many such databases are commercially available, such as the popular dBase III and Lotus 1,2,3. Only the minority are specifically tailored for the user's needs. The majority require the users to tailor their needs to these existing "generic" databases. What they have in common is that they are either hierarchical, network, or relational in their structure (some claim to be a combination of two or more). These three make up the vast majority of the internal architecture we refer to.

When we talk about internal architecture, we are describing how the company is going to use the data stored in its common database. Will it be rigid by rote combinations, such as calculating interest payments? Or will it be flexible to a degree that most combinations and permutations possible are known beforehand. An example would be a choice of options available to a purchaser of an automobile. Or will it be extremely flexible such as the travel destinations of airline passengers? Here, the combinations and permutations can become very large and, hence, not totally pre-known. Each of these three examples represents the application of a different database architecture; that is, hierarchical, network, and relational, respectively.

The goal of the internal architecture database (commonly referred to as the database) is to give the user the ability to use the data for his or her particular need as spontaneously and as completely as possible. We are then interested in how the data are accessed, organized, recalled, and used. This means a successful database has to comply with the following criteria.

1. The data have to be available when needed.
2. The data have to be complete, not fragmented.
3. The data have to be organized in a recognizable format.
4. The data have to be accessible to users without need for intermediaries.

These are the minimum conditions necessary for general use. It is interesting to note that we have been striving to achieve these minimum standards for all of recorded history, that is, where information has had to be received, acted on, and then sent on its way to the next recipient. This is not a new criteria discovered for CIM applications.

All three internal architecture databases meet these standards. They each have their particular uses where they are superior or at least equal to the others. They are all meant to be replacements for slow, difficult to access manual databases. This is easy to illustrate. A typical manual industrial database contains

- Policies governing how the firm operates
- Engineering drawings
- Engineering, manufacturing, finance, marketing, employee relations, etc., instructions
- Reference books
- Letters defining each of the above
- Data collected and recorded on multiple types and numbers of forms
- Files in numerous locations: libraries, office files, job folders, etc.

The byword is, "It is all there but slow and difficult to find." This means the databases are not used as extensively as they ought to be. We all know that wherever possible we will wing it and depend on our not-so-superior memories or experience to solve problems because it is too hard to find and use the data. In a world where we have to respond quickly to situations, this is not satisfactory, and has led to the internal architecture database development.

Computer databases contain the same information as listed before, the difference being that computer databases are structured for easier access. It is the ease of access that we strive to improve. This has to do with the design of the database (how it works within the CPU of the computer) and how easy it is for the average user to work with. Let us look at the common concern that has to be dealt with in conceiving database architecture and then the pros and cons of the three generic types.

Databases should be easy to use in that the data can be retrieved easily. Is this so? The answer is yes and no. Yes for those who are computer literate, no for those who are not. Computer literacy has many variations: from those who know how to turn on the computer and follow simple prompting instructions to those who are proficient systems analysts and programmers. The goal of most designers of CIM systems is to serve the needs of the former. There are several ways of getting around the literacy problem. The choices are

1. Train all users.
2. Train key users only.
3. Establish a priest and priestess class to handle the computer interfaces with the users.
4. Aim for less need for computer literacy.

Choices 1 and 2 require the establishment of a training infrastructure. This means considerable costs and, like all such bureaucracies, tend to become intolerably inefficient.

Choice 3 is the antithesis of CIM. We want the computer and its communicating capabilities to be available for all to establish communications excellence within the organization.

This leaves choice 4 as the only viable alternative. Indeed, this is the path virtually all computer hardware and software firms are tending toward. In a CIM environment we want all manner of people doing a wide variety of functions to have access to the common database, to use the information and to make the required updates as the situation dictates.

In business applications we have experienced three types of internal architecture databases: hierarchical, network, and relational. I will explain each.

The Hierarchical Database This database is the first to be employed. In many ways it is an attempt to parallel the human organization chart for information flow. The database is set up to emulate the way the designers of the organization intended for it to carry out the assigned mission of the specific function(s).

Note the fixed lines of communications implied in the organization chart of Figure 3.1. This is nothing more than the familiar pecking order chart employed by virtually all businesses to demonstrate the relative rankings of the various functions. It would imply that each grouping is an isolated entity transmitting information up and down the chain of command with no input from parallel organizations or even between subfunctions reporting to the same function. While the chart may adequately portray the organization, it is a poor model for an information flow system.

This database does not represent a suitable vehicle for successful CIM implementa-

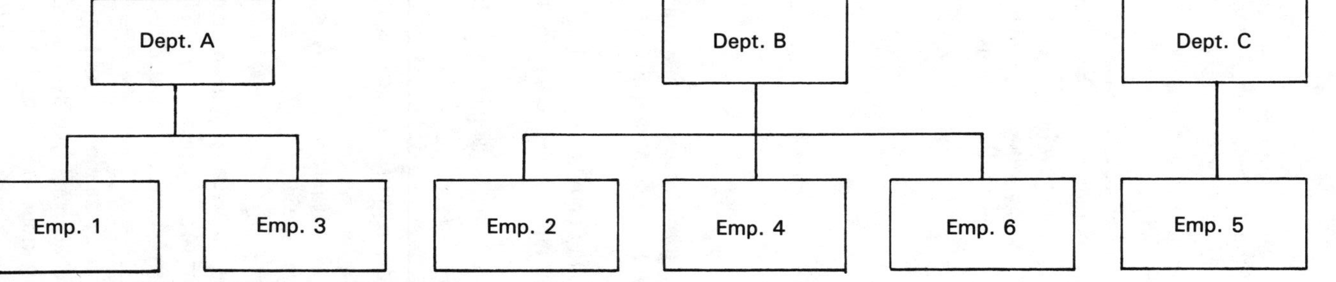

Figure 3.1 Hierarchical database (rigid structure).

tion because it is too rigid. Designers following this model find it too rigid. Information flow does not follow organization flow. Hence, hierarchical databases are not used for CIM. If we followed this model to its logical conclusion, we could not have electronic crossing of data between departments. Neither is there any rationale that would instruct us to preplan "what if" scenarios for comparing parallel sets of data.

Let us look at an example of why the hierarchical database is not a common database (a requirement of CIM). Suppose department A and department B of Figure 3.1 represent the quality control department and the design engineering department of a manufacturing firm, respectively. By data collection techniques, the quality control department can input process defects occurring, with their causes, into its database. Similarly, the design engineering department can have defined and calculated all overstressed areas of the designed structure being manufactured. They even may have it all in their working databases. Now, suppose a quality engineer and a design engineer decide they can make improvements but only if they can correlate defect types and occurrences with highly stressed areas of the design. How can this be accomplished? With a hierarchical database, not electronically. It will have to be a manual comparison because there is no data linkage between the two departments as the database is designed. So we can see that the hierarchical database is not a true common database, as required for CIM. It is a stand-alone database that predates the concepts of CIM. It is analogous to "islands of automation."

The Network Database This is currently the most frequently used of the internal architecture databases. It is an attempt to correct the deficiencies of the hierarchical database. The network database recognizes that information flow is not similar in nature to organization structure. In fact, information flow is much more complex and convoluted. Figure 3.2 represents the network database.

Note how convoluted the flows are in the figure. We have employees and projects and other entities all linked directly or indirectly (the dashed lines). There is no pretense of structure along organizational lines. In fact, there would appear to be no pretense to any form of organization, but there is! It is organized the way information is thought to flow within the subject function(s).

The internal architecture database patterned on the network philosophy takes advantage of the inherent speed of the computer. By doing so, it can construct direct paths between potential communicators. This makes it a very efficient user of computer capacity as well as being very fast.

The success of this type of database presupposes that all potential, or probable, combinations and permutations of data uses can be defined. Therefore, it is possible to write computer programs (called subroutines) that link all the preconceived possibilities with most efficient paths. The database is structured to look at certain addresses within the computer memory for files containing information that by program is linked to other preordained combinations.

By using the design engineer and the quality engineer example again, if we want to compare process defects with overstressed design areas, then the designer of the network would have to know that this is a likely possible comparison area. Hence, a network would have to be predesigned for the two database components to be able to

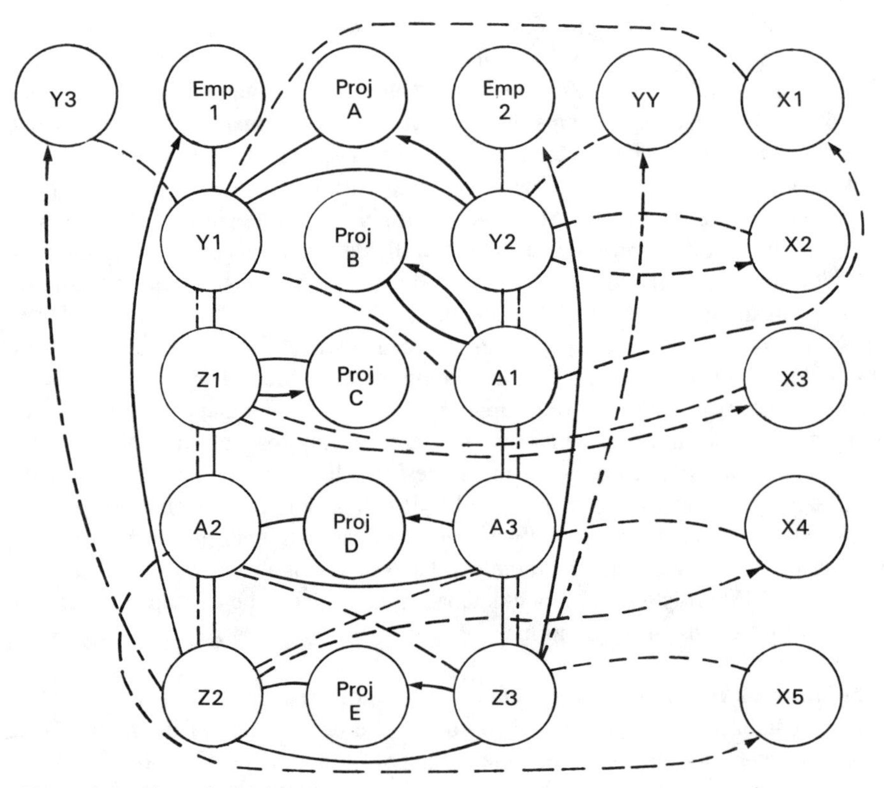

Figure 3.2 Network database.

communicate. This is exactly what happens. The designer goes to great lengths to get the user community to list all the possible "what if" scenarios that should be designed into the database. If this exercise is successful, then a networking database is like magic. It appears to be capable of answering virtually all the questions put to it by the users. It is the best of all the databases in terms of response times where all combinations and permutations can be predetermined. Even if all is not known, all is not lost. It is still possible to restructure the program to allow additional combinations and permutations but, obviously, not in real time.

In the example just used, if the comparison possibility had been preconceived and structured into the network, then the comparison of defects to design stress could have been accomplished very quickly. If not, it still could have been done electronically but only after a new subroutine had been written and tested. Sometimes writing the addition to the program can pose major problems. This depends on the complexity of the network and the degree of change required to accomplish the new request. Usually, subroutines are not too difficult to accomplish. With many of the newer networking databases, the authors have tried to make subroutine writing as simple as possible so as to give as close to real-time capabilities as possible.

The networking database is a true CIM-compatible database. It is conceived on the idea of interchange of information between realistically determined users. That being the case, it does champion communications excellence.

The Relational Database This is the latest entry into internal architecture database design. Its purpose is to give the user the most flexibility possible in using information stored in the common database. The network database, while remarkable in its usability, does have the limitation of requiring predesigned combinations or paths for data to follow. Many times, especially in manufacturing activities, it is very difficult if not impossible to predetermine all the comparison of data activities that might occur or be desirable. With the network database, the only solution is to write a new subroutine, which may not be practical. The lack of ability to make ad hoc comparisons is particularly irksome when trying to solve process problems that will not allow the shutdown of the process while achieving a suitable solution. This is particularly true in flow manufacturing situations, where shutdown often creates worse problems than living with the lesser difficulties. To improve common database performance in the real-time mode, relational databases were developed.

Relational databases are the most flexible of the three discussed here (see Figure 3.3). The philosophy of its design is to completely ignore the paths along which data will flow. The need to determine combinations and permutations of possible use is eliminated. Instead, all data are treated equally. Data are assembled in tabular form as it is entered. All that is necessary is that it is given a descriptive identification for indexing and recall purposes. The programming effort is primarily directed at making it easy for users to find tables that are of interest to them. The database consists of tables of information entered at random and a series of commands that allow users to make comparisons as the fancy strikes them. Therefore, it can be entirely ad hoc, something not achievable with the network database. Programs written in conjunction with a relational common database tend to be algorithms that enhance the user's capability of making data comparisons. As one can imagine, this does lead to a very flexible, almost unlimited, manipulation of data to solve the user's needs.

Let us look at an example of using a relational database. If our hypothetical quality

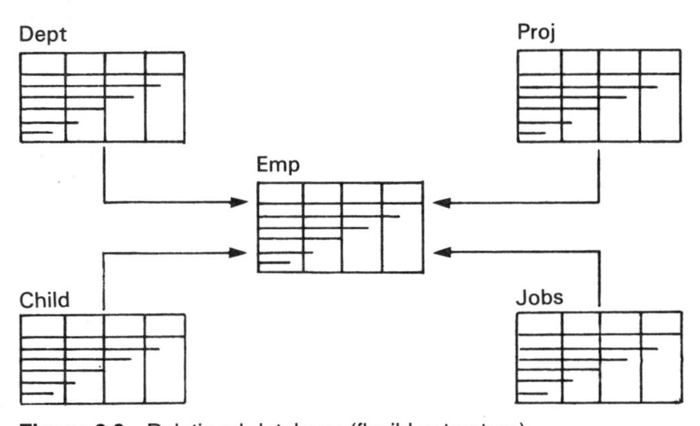

Figure 3.3 Relational database (flexible structure).

engineer and design engineer thought there might be some connection between process defects and areas of the design where the stresses may have been over an arbitrary limit, it is easy to check this out. In the hierarchical case, it could only be done manually, which probably means it would not be done. In the network case, it could not be done directly (hence, quickly) since it would be unlikely that such a desire would have been preplanned. The network would require a subroutine be written to achieve the desired results. This, too, may not get done, primarily because the time to do it may be too long. However, with the relational common database, it becomes a simple task of having the computer make two lists and then compare the list electronically. Generically, these are the commands given to the computer.

- List all causes of process defects.
- List all areas of overstressed design.
- Compare defects with overstressed areas to see if a correlation exists.

And that is it. Since the database is structured only in that it lists data in defined tables, it will now search through all the tables to see if any process defects are included, and then it will do the same for overstressed areas. It will print out these two found data sorts. Then the program will direct it to create a new table containing matches of defects and overstressed areas. This becomes a new table in the common database and also the answer the two engineers sought.

By using the logic just described, the relational database creates new data tables ad infinitum to satisfy new comparison requirements. Because of this unstructured approach, relational databases are slower than network databases, sometimes unacceptably slower. I have heard of cases of early relational databases taking hours, sometimes more than twenty four, to conclude a search. Relational is definitely less efficient than network but is really within the spirit of the CIM philosophy. It allows for true interaction of information between functions on as close to a real-time basis as practical. Much work is being done right now by computer hardware and software vendors to merge the speed and efficiency of network databases with the flexibility and user friendliness of relational databases. Many researchers believe the inevitable maturing of relational databases will lead to the real revolution in business management and control; that is, the ability to manage in a real-time dynamic mode.

Now that we have described the three common internal architecture databases, which one is best for a business considering a CIM solution? There is no clearcut answer. We can state with a high degree of certainty that hierarchical is not compatible with a total CIM strategy. This leaves network and relational, which are both compatible with a total CIM strategy.

Before leaving hierarchical, it is important to reiterate why it is not suitable for CIM. Hierarchical is modeled after the typical rigid organization structure. It allows data transmissions only along a chain-of-command flow path, which is not how we really communicate. We all know that the informal information flow process (the rumor mill, the water cooler telegraph, etc.) is faster and frequently as accurate as the official information channels. This is true because we do talk to and meet with contemporaries from other functions, and we do conduct transactions with them even though the official channels depict the flow being up the line to a common head and then radiated downward again. Official communication strategies recognize vertical dissemination of information but hardly ever recognize horizontal dissemination of information. Most organizations, by necessity, depend on the informal information flow for them to exist in a state of acceptable efficiency. The hierarchical database does not recognize this informal flow, and, hence, is flawed from the very beginning. Since CIM depends on both vertical and horizontal dissemination of information, a hierarchical database structure is inadequate.

Using a network or a relational database depends on the circumstance of the organization desiring to use a CIM solution. If there is little deviation from preconceived planning (possible combinations of data necessary to carry out the function's goals), then a network internal architecture database is sufficient, probably preferable. If this is not the case, then relational is required. This is especially true if preconceived plans are meant to be guidelines only.

Unfortunately, most manufacturing businesses use preconceived plans as guidelines, so network databases are not totally satisfactory. This is so because more than 70% of all product producing firms fall into the job shop category. These firms rarely have detailed specific production plans beyond three months, and, hence, their preconceived plans are more strategic, with contingencies in varying degrees of sophistication, than preplanned in nature. For this reason we can see that relational structured internal architecture databases would be more suitable for CIM for the majority of manufacturing businesses.

The CIM database selected has to meet four criteria.

1. Data available when needed
2. Data complete, not fragmented
3. Data in a recognizable format
4. Users have easy access to the data

Table 3.1 compares the database choices against these four criteria for typical job shop manufacturing entities. Relational comes out best, but if the firm's preconceived planning is very accurate and complete, network would probably be an equal alternative.

We have now completed our explanations of internal database architecture, and now it is appropriate to introduce external architecture databases.

External architecture databases relate to how data are input to the computer, that is, how data are manipulated for multiple-user applications. Recall that internal

Table 3.1 Database Architecture for Manufacturing

Data have to be	Manual Database	Hierarchical Database	Network Database	Relational Database
Available when needed	poor	fair	good	excellent
Complete, not fragmented	poor	good	excellent	excellent
Organized in recognizable format	poor	poor	fair	good
Easily accessible to users	poor	poor	good	excellent

architecture databases had to do with how the computer uses data to do the work required by the program. It is internal to the memory core of the computer. Think of it as the brain while the external architecture database is the arteries and veins feeding the entity. External architecture databases are called by many the essence of CIM. This is so because CIM is communications excellence, and a well-thought-out external database system makes communications excellence possible.

Once the criteria for a CIM system are known (e.g., what components of communications and control are required for the business and who and how components have to be linked), the external architecture database can be selected. There are three primary types of external architecture databases. Each can be structured to be used with the three internal architecture databases. These three are called independent, centralized, and interlinked. In the CIM concept, these are means of organizing data and the paths by which these data flow so that multiple users from different disciplines within the company can all use the data as they need. An example of data flow would be as follows.

Suppose the engineering department has collected data pertaining to load deflections of axle shafts with different imposed loads. They use this information for design analysis, and they keep these data on a computer system storage disk. Now, suppose the manufacturing department is required to machine the shaft such that it has no deflection under the steady state load. This means that the shaft has to be machined on a bias. An offset has to be machined into the shaft so that when it is under load it will deflect to a straight and level position. The manufacturing department can go about gathering deflection data such as the engineering department did and then calculate the offset. In a non-CIM situation, this may indeed be the way it would be done. But that is waste! The best thing to do is to use already existing data. If the external architecture database is a CIM type, then it can be done quite easily. It is a simple matter for the pathways to the data to be set up so the user (manufacturing) can access it, merge it with its own data, and come up with the deflection solution it needs (manipulated by the internal architecture database). This is exactly what occurs over and over again through the structure of the external architecture database. Data flow is established so all users can get at it easily on an as-needed basis. The original home base of the data becomes irrelevant. What is important is that the originator of the data has placed it into a system that allows others to use it. This is called the "common database" and is the basis of CIM and the reason for our interest in external architecture database systems.

As was just stated, there are three primary types of external architecture databases. Let us look at them specifically as to what are their pros and cons concerned with making the CIM common database concept work.

The first one to consider is the "independent database" (see Figure 3.4). This is the oldest approach, going back to the early days of computers when data were organized for specific users. When computers were not prime communication tools, little thought was given to the potential needs of other organizations. The users were primarily concerned with satisfying their specific needs and little else. Prior to the CIM concept development integration (e.g., structuring of databases to a universal standard for all users), it was not considered important. Therefore, external databases were entered

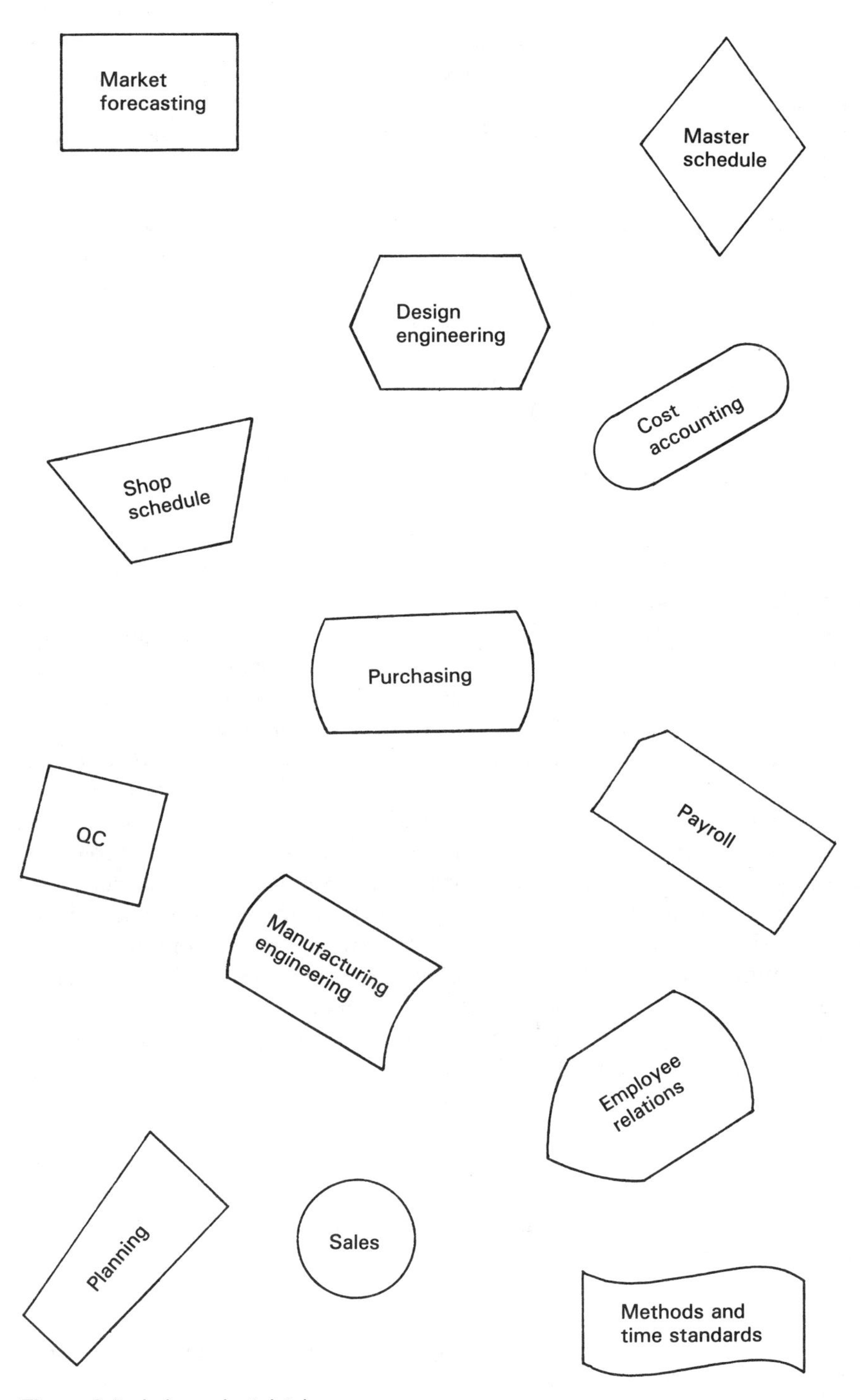

Figure 3.4 Independent databases.

into the computer in a manner that we would now consider to be haphazard. Therefore, use of independent databases is not considered to be a good model for CIM, although it can be made to work.

There are several reasons why independent databases are not considered satisfactory for CIM.

• Since the databases are not necessarily similar in structure, there may be extreme difficulty in transferring information (data points) from one user's application program to another. The syntax used to structure data and how data are to be used will inevitably be different, thus requiring a complex translating device to be invented and employed. Even worse, the databases may be written in different languages, making the translating task that much more difficult. This is a severe detriment to the desire to share data between business functions.

• If a group of the users decide to update the data structure to improve the efficiency of their uses, it may prove to be a severe blow to other organizations tapping in to get the data for other uses. This typically happens because the "owner" of the database is not considering the needs of outsiders as he or she optimizes (usually suboptimizing and thus deoptimizing the overall effectiveness of the entire entity). So, a change in the external database structure requires changes in all other external database structures still wanting access to the contained information for their own uses. This normally means high cost, confusion, and extended time before the system settles down again to an acceptable situation.

• Usually with independent databases, redundant files have to be maintained by all users. Since the organization of the independent databases is usually different, the method of information transmission is usually through flat file transfer. This means transfer through a translator-type program. Translator-type programs tend to be complex, negating the option of the real-time transfer simply because they are complex. Hence, a "data dump" occurs on a scheduled basis, requiring the maintenance of a redundant file by secondary users. An example, and one that is all too common, would be the geometric model file devised by the engineering department and used also by the manufacturing department. The danger of not having real-time capability is that the primary user may make a change affecting the secondary user who would be oblivious to it for a period of time. That dwell time could be critical in some circumstances.

• The cost of coordinating independent databases is usually excessive. This, along with the need of a superior auxiliary communications system to tell all users that a change is occurring, makes it cumbersome and difficult to deal with. Independent databases require multiple program changes whenever the primary user decides to upgrade the system, making it obvious that the cost is going to be beneficial only to software programming agencies. High costs and the need for intensive management to make this type of common database work are significant disadvantages for independent databases.

To give a balanced perspective, we should point out the inherent advantage of independent databases. When structuring an external database system, the independent database knows no equal for tailoring a data flow for specific use. Therefore, if a use can be discerned where there is low probability that a secondary user will emerge, then it is proper to tailor the database for that single user. In such a situation there will be no need to compromise between organizations for different needs. An example would be in certain subfunctions of engineering. Finite-element stress analysis is used only by the design engineers and very rarely is employed by other functions, and hence, it is a good candidate for independent database use.

The next development in external databases was the "centralized database" (see Figure 3.5).

Once the unifying concept of CIM was recognized (probably through the desire to link CAD and CAM), the need to share data efficiently between business functions became important. It became evident that making it easy to use shared data provided significant productivity improvement. To gain this advantage, obtaining a real-time information flow becomes necessary. This led to the search for ways of organizing data in a manner that was suitable for dynamic, rapid information flow. The first successful attempt was the development of the centralized database.

The basic characteristics of this type of external database follow:

1. All data are generated in one format, regardless of how they will be used.
2. All data are stored together in one massive data file.
3. All data are processed by the same program or by sets of companion programs.

This system ensures good communication because the whole equals the sum of the parts at all times. There is only one program that does all functions the business deems necessary in the avenue of communication and actions/reactions. However, there are problems of an opposite nature to those of the independent database with this approach.

1. The administration of such a system is unwieldy for even a moderate sized company. The external database rapidly becomes very large, and, as it becomes larger, the ability to maintain universality becomes more difficult. This means more compromises have to be made so that the data can be accessed by more and more diverse users.

2. Access time to retrieve specific information may be too long. Recognizing that what may constitute being too long is very subjective, this still has to be considered a significant fault. Engineers may be satisfied with a delay of minutes for a response to calculations. Operators at factory work stations may complain about delays of seconds. Obviously, response times are related to perceived norms. However, it has to be recognized that a very large multiple purpose program is bound to be slower than independent programs running on stand-alone systems. Hence, the complaint that a centralized database would tend to get to be objectionably slower is a valid one. Mitigating this encroaching slowness would be the selection of and upgrade to larger and faster computers.

3. Maintenance of a large centralized database is very complex. This tends to evolve into creating an organization to manage the database. Many times this is the antithesis of CIM. In CIM we want to create spontaneous responses to stimuli and to allow creativity of expression and management solutions to flourish. CIM requires communication excellence. Many times, with a large maintenance organization created

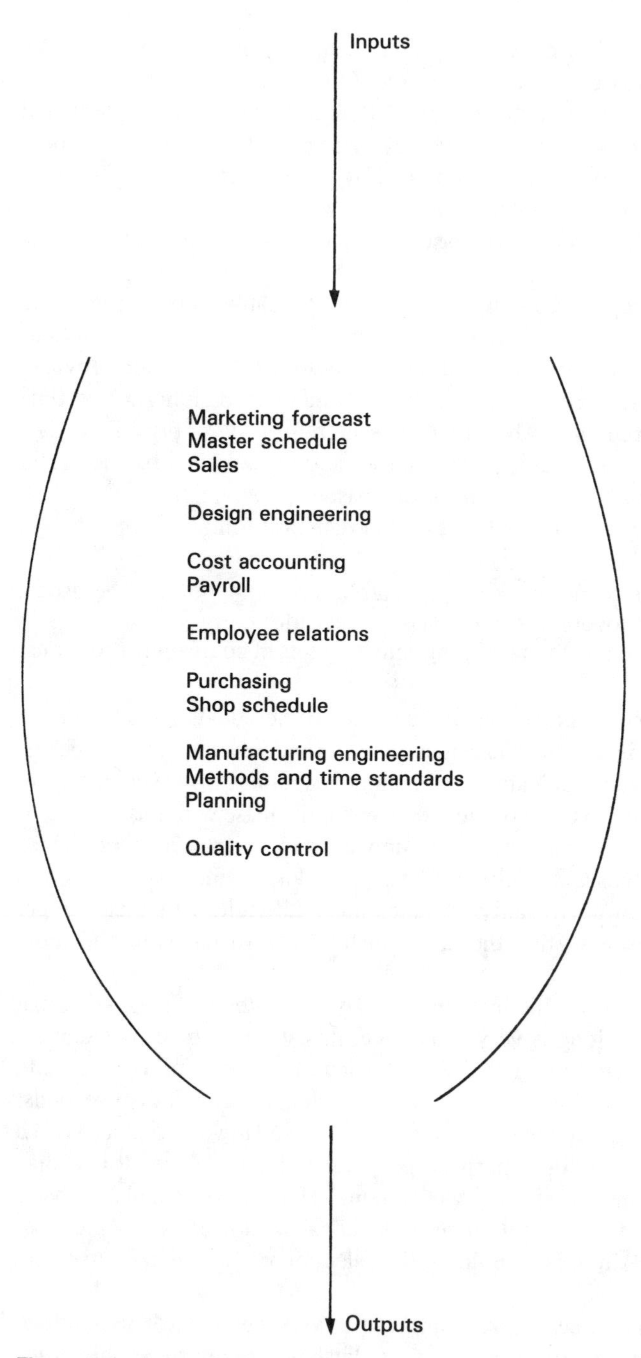

Figure 3.5 Centralized database.

to tend to the needs of the database, this creates a buffer between those who need to use the system and those who administer it. Unfortunately, this too often manifests itself in a self-righteous bureaucracy that does what it conceives to be necessary to maintain the efficiency of the system and not the efficiency of communication transfer and ease of access of the users.

4. Since the system is a centralized one, a large amount of the information stored in the database may be of interest to only a few or a limited number of users. This may mean response time is inefficient because the database may have to sort through voluminous files of irrelevant (to the current user) data before finding and manipulating the data the user is interested in.

The inherent benefits of a centralized database are derived from the philosophy employed in its basic structure. In many ways the benefits are the counterarguments to its deficiencies. The usually agreed upon benefits follow:

1. All the information that a company has pertaining to any specific and or related activities are stored in one data file. Therefore, there is little difficulty perusing the library for all the pertinent information. Research is easily accomplished.

2. All users within the company can have access to all information (unless restricted for security reasons) on as close to a real-time situation as is practical. There is also no need to be concerned about how current the information is because updates made by the prime users and data handlers are immediately available to all who are linked to the common database.

The last and most current form of external database used for the common database is the interlinked database. This type of database is a result of experience trying to make the previous two work. There are many variations of the interlinked databases. Our intent here is to describe them in a generic sense and to leave the investigation as to what variation suits a particular user best to managers investigating specific tasks of matching their organization's needs with a multitude of commercial offerings.

This concept (see Figure 3.6) requires all the functions of an organization to develop and maintain their specific external database in accordance with a set of rules, sort of like city-land development zoning ordinances. We have an interlinked database when data are entered only once, and it becomes available practically in real time to all other functions who need to use the data. The data are distributed in real time to all other users, typically on an as-needed basis. This keeps data flow to a minimum and enhances efficiency. Each user has control over their database, but the data can be requested by and loaded into other user's databases and vice versa.

In this concept, common data essential to many different users are kept in a common master file, and specific information needed by a single or a very few users are kept in local files so that there is a definite mix of the independent and centralized concepts. Data are independent where it makes sense for them to be, and are centralized where economy of scale and user common needs dictate. This merging of both concepts allows us to design specific databases for individual company needs. The basic rules for developing interlinked databases are few but very important. The rules have to be compatible with the precepts of CIM (e.g., create communication excellence through the use of easily accessed common databases). My version of the rules follows:

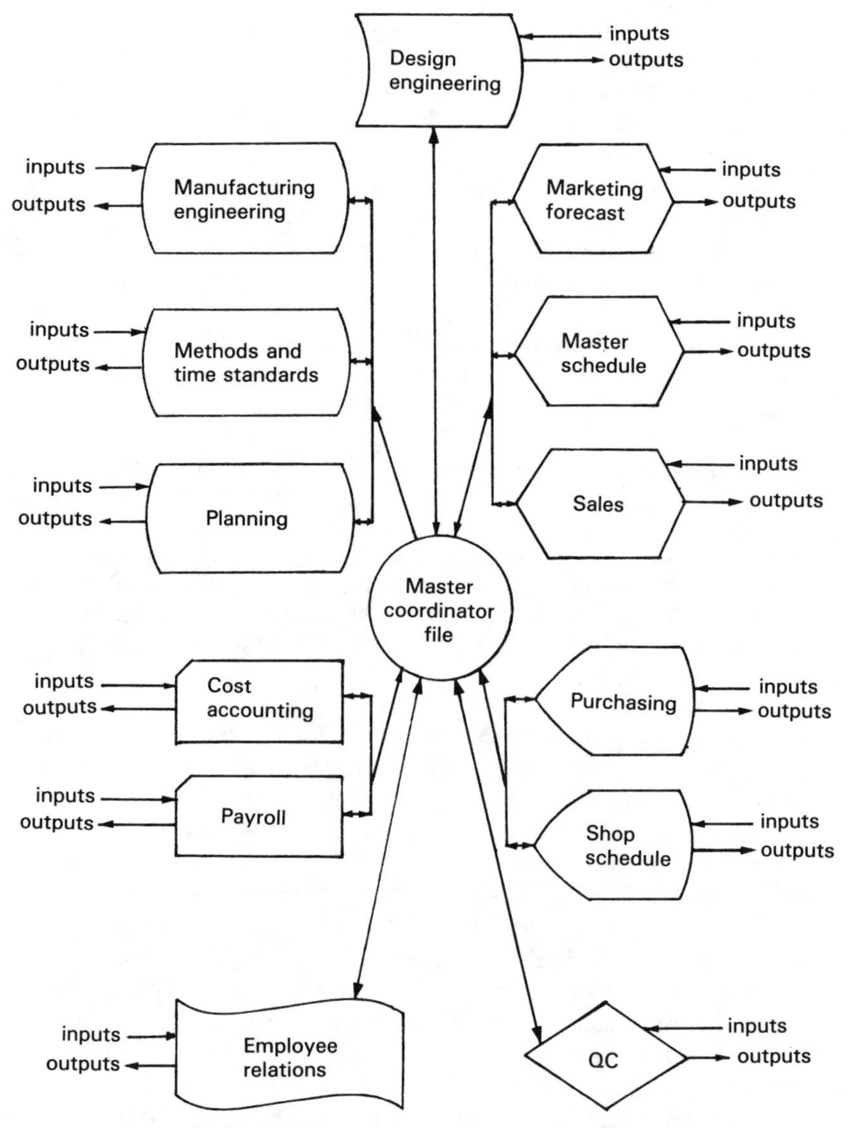

Figure 3.6 Interlinked databases.

1. Establish only one type of internal and external architecture for the entire organization.

2. Write all applications programs in the same language.

3. Do a detailed analysis to determine which data are mostly common and which are only occasionally common to multiple users, and then make the master file as limited as possible.

4. Develop as few rules as possible concerning data structure and input. Select a database that offers as much ad hoc capability as is practical.

5. Let users control their own databases as much as possible. Minimize software maintenance personnel as much as possible.

Chapter 4

Control of Business Operations by Means of CIM

So far, we have looked at databases, both the internal and external architecture, that can be placed on computers. But how are all these data that we now know can be linked together used for an entire company's operation? As managers we are interested in creating information flows so the business runs smoothly, is efficient, and can react to outside stimuli quickly as conditions change. We want the magic of instantaneous information, not too much and not too little, available to all the functions of the business. We want information flows that are just right for our specific needs. The way this can be accomplished is through proper structuring of information flows. This structuring is necessary regardless of whether or not the system is "computerized." Therefore, whatever is right for good communication structuring is a required precursor for a system that also uses the benefits of computers.

There are many ways to structure information flows. However, since we are mainly interested in industrial organizations, let us use the classic seven steps of the Manufacturing System (see Table 2.1) as an example of flow structure. Here, we see a logical sequence of information from customer order entry to shipment of the completed product to the customer. We also see scheduling and monitoring activities. There is a concurrent dialogue necessary with assorted vendors. There also are distinct side-bar loops such as information from engineering to manufacturing and back again. From analysis of this information system one thing can be discerned. Not everyone has a requirement for the entire set of information contained in the system. Therefore we can see that a structure of an information system is logically based on a need-to-know format. Pause and think and you will surely agree that the need-to-know format for information distribution is fundamental. Otherwise, we would be buried with data that have little value to us, and the valuable data would be hard to decipher in all the clutter. If the computer is used to enhance the communication flow, then it must be used in a manner compatible with the need-to-know concept.

How, then, do we use computers to enhance the information system? We have seen in chapter 3 how databases can be structured to establish the common database. We know that centralized and interlinked external architecture systems are powerful systems for communicating commonly desired information between the functions. But how is all this hooked together to give us a physical reality? That depends on the size of the organization. For example, the organization can have the data on one computer with a versatile and probably large core storage and memory system. Or, it can have many independent computers, each with a piece or several pieces of the data (but this would not be a CIM solution, so we can ignore it). The data can be resident in segments on many computers linked together. Each of these linked computers represents a specific function and, hence, the need-to-know philosophy applies. From practical experience we have learned that the single computer solution is not satisfactory for any but small organizations. This leaves us with the linked computer system as the most practical manner of using computers with information flow structures.

The need-to-know philosophy tends to create a hierarchical ranking. This is logical. By using the seven steps of the Manufacturing System as our example, it is logical that the manager of the organization would have a requirement to be aware of current order rates, shipments, basic concepts of the design being employed in the products, schedules on a broad-based status level, and other items necessary to represent the company to outsiders. Likewise, it is logical for the engineering manager to have a detailed level of information concerning product design, schedules, and designs being employed in current production, but very little need for other information to do his or her job effectively. Therefore, we can see that the type and completeness of information changes with the level of the organization's management structure. Therefore, the most common structure of linked computers is the hierarchical structure. Here, each computer deals with the subset of information required for the function for which it is providing communication services.

Recall that the hierarchical internal architecture database software was not considered a good model for human organization nor data flow organization because of the many parallel and sideways information flows found in ordinary information discourse. Then, why is hierarchical sufficient for computer linking strategies? The answer is straightforward. With hierarchical it is possible to design the various computers in the linked systems to be specialists in their assigned tasks (more on this later); therefore, they can be very efficient at doing specific things. Also, even though they are placed in a ranked order, nothing prevents them from being linked sideways and parallel even though the primary information flow for performing work is up and down. Besides, computer ranking is meaningless on human terms. They are machines designed to perform certain tasks. As always, machines are laid out to optimize a work sequence as well as possible. Remember, we are considering how hardware will be structured to carry out the data flows dictated by the database software and not the degree of flexibility of the software. We will still have the ability to use a relational structured common database. In fact, it will work better with a hierarchical-linked computer network because the components will be better able to handle directed data.

The main reason for hierarchical is distribution of tasks in accordance with the need-to-know philosophy and division of risks if a computer module should fail. Figure 4.1 demonstrates the assignment of tasks to various special types of computers and visually displays that one down unit does not mean the entire system is not functional. If we had only one computer doing everything, then we would be highly vulnerable to shutting down the entire enterprise if one piece of equipment failed. If the system crashed (computer slang for the system failing for whatever reason), then all users are very likely to lose their current data. This could be very costly for many manufacturing processes and, hence, there is a strong desire to have redundancy. There are two ways to have redundancy: have a duplicate central system (very costly) or distribute the risk by means of linked special purpose computers with duplication on key items only.

Another very cogent reason for adopting the hierarchical-linked computer concept is the cost of future expansion or change. With a central system, change and expansion is more difficult. Since one computer is running one common database with perhaps minor database subset segregation allowed by the large computer's design, change is difficult owing to its inherent complexity. Also, since the cost of the central system is bound to be more than individual components of the linked system, starting small and then expanding is not a likely option. With a linked system we can start out with simple tasks, such as time and attendance. Then, after gaining experience and value from this information, we could decide to expand into full data collection, probably by adding more devices. With the centralized concept, if data collection was to be a future possibility, then it would have to have had the capability built in at the very beginning, at the commensurate extra costs. A large central computer employed for multiple tasks has to be more rigid than an assembled linked series of smaller computers.

With the advent of minicomputers and microcomputers, the hierarchical concept became more practical. With the networking concept now becoming more advantageous, especially as standards for communication (such as MAP/TOP) are accepted by different hardware producers, hierarchical becomes the only logical choice. One computer can be assigned one primary task, and that computer can be designed for that task. A good example of this concept is the Digital Equipment Corporation's venerable PDP-11 minicomputer. The DEC PDP-11 was designed specifically for engineering design tasks. It allowed engineers to do finite element stress analysis because it could do very large numerical calculations. But to do this it sacrificed speed (e.g., it had slow transaction rates). Here is an example of a special purpose design, and in our hierarchical linked system it would take its place in the design engineering function. No one would consider using the PDP-11 for general purpose computing (payroll, scheduling, word processing), but one would consider using an IBM System 36 for that. Two computers, used for two different applications. Each computer is matched to the proper application.

Computers in a hierarchical-linked system have to be matched to their assigned tasks. What are the guidelines that can be used for this matching purpose? There are some generally agreed on concepts for hierarchical linking of computers in a CIM system.

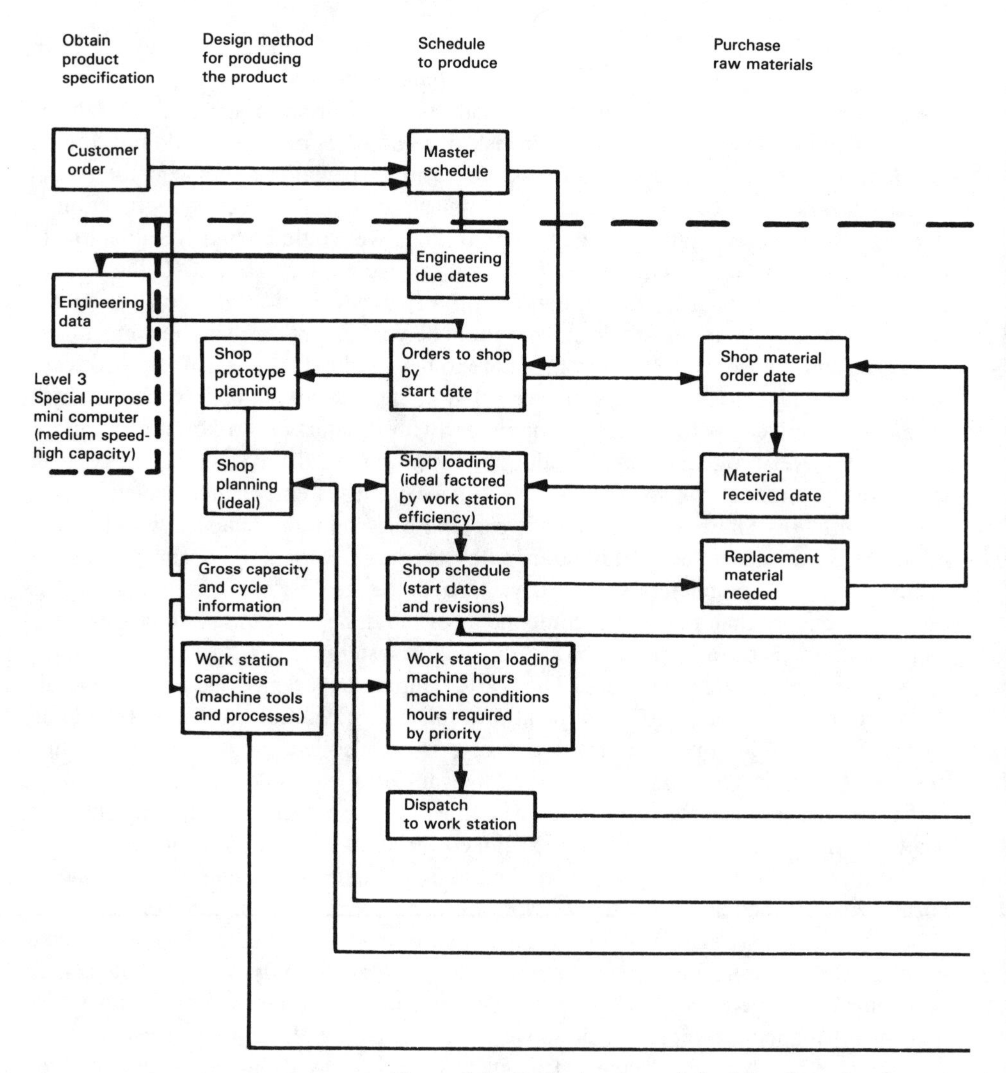

Figure 4.1 Typical variant of the Manufacturing System

• The higher the level of the computer, the larger the part of the manufacturing system it controls. Higher management levels supervise more diverse activities than lower level managers.

• Reaction times are slower as we go up the hierarchy, analogous to a general manager deciding on a business strategy who usually takes longer than the decision time a foreman takes to assign work to a machine station.

• Real-time requirements decrease as we go up the computer hierarchy. A decision to accept or reject a part at an on-line inspection station has to be made in seconds, while the decision to place an order with a vendor may require a minimum time of hours.

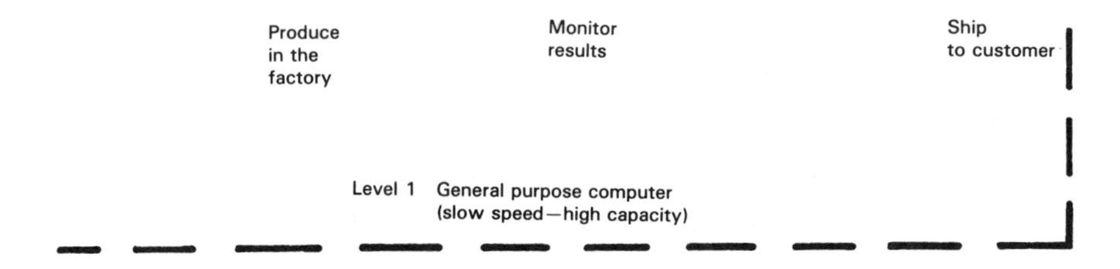

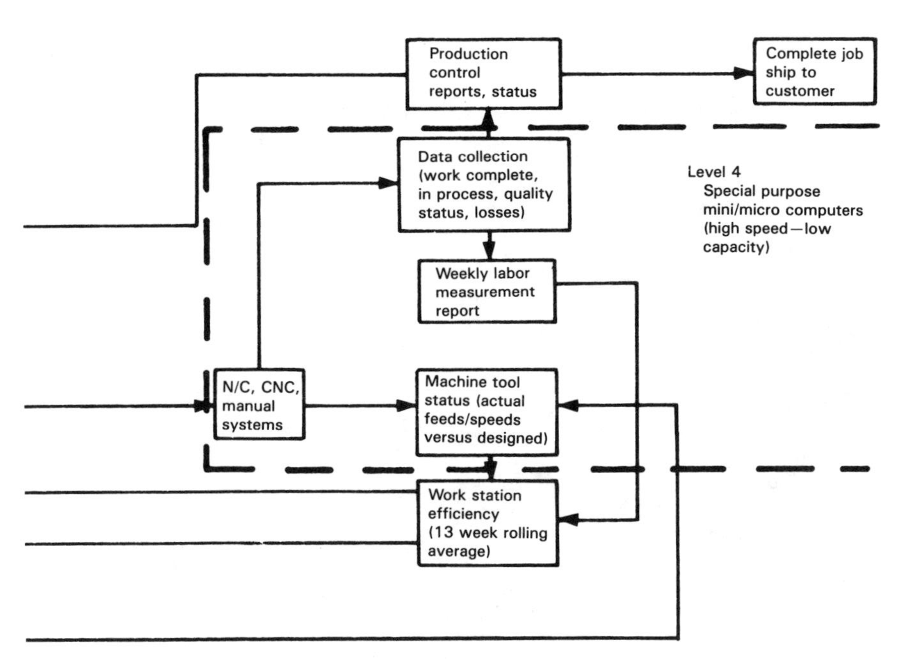

with hierarchy linked computer flow superimposed.

• The speed of information slows down the higher on the hierarchy a computer is. More deliberation is required at higher levels; hence, information flow rate is not very important. Completeness of information is more critical.

• The higher the level on the hierarchy, the more general purpose and less specialized the computer has to be. The general manager, for example, has to have summary style information on many topics but very specific details on few topics. Therefore, the supporting computer has to do many tasks adequately but none at the superior level.

Now that we have described the structure of computers desirable for CIM, let us look at what that software and hardware system will be working with. This is most often referred to as the common database. The common database is different for every company, but there are some generic contents that can be considered universal. Another way of saying it is that there are no set definitions as to what constitutes an all-inclusive CIM database (another way of saying common database). Every type of organization will probably have a different type of common database. For example, a database that could be said to be representative of discrete parts job shops will be discussed next.

This type of database could be resident on any combination of internal and external architecture systems previously described. It would be up to the systems engineer to decide the details as to how the common database works with the chosen architecture. Such things as speed, accessibility, and flexibility are considered (see Table 4.1).

The eight categories are basic for job shop manufacturing. If we consider the "two knows" of manufacturing, often considered the basic tenets of manufacturing, we see that the primary requirements to produce the product are all there. We can see, then, that these are the fewest numbers of databases that can be grouped to form the common database. For illustrative purposes the databases are grouped under their respective "know."

Table 4.1 The Common Database for CIM

The Basic Contents

1. Design characteristics
    (1) Geometry
    (2) Material
2. N/C tool paths
    (1) Interpretation of geometry for machine tool movements through space
3. Planning
    (1) Methods
    (2) Process selection
    (3) Time standards
4. Purchasing
    (1) Definition of parts to be purchased
    (2) Stocking plans (inventory control)
    (3) Vendor contract requirements
5. Quality requirements
    (1) Quality plan
    (2) Quality historical records
6. Master production schedule
    (1) Dispatch activities
    (2) Purchased and produced material due dates
    (3) All schedules
7. Data collection
    (1) All feedback on activities
    (2) Compares plan to actual (nonfinancial)
8. Cost analysis
    (1) All costs from other database inputs
    (2) Compare financial plan with actual

Know How To Make the Product

Design characteristics: The basic form and function of the product.

N/C tool paths: Digital instructions to the process-performing equipment on how to shape the product.

Planning: Manual instructions to the operators of the process-performing equipment on how to shape the product.

Purchasing: Based on the design characteristics and planning, instructions for buying nonproduced parts.

Quality requirements: Vendor, in-process, and customer acceptance requirements necessary for successful transfer of title at job completion.

Know How Long It Should Take To Make the Product

Planning: The calculation of time to perform the required operations to fabricate, test, and ship the product.

Master production schedule: The analysis, logistics, and coordinating of the designated planned times to make the product in accordance with commitments to customers.

Data collection: The gathering of information to be used to compare planned (budgeted activities) with actual performance.

Cost analysis: The conversion of times, planned and actual, to dollar costs plus the addition of purchased parts costs, planned versus actual, to evaluate financial status of the operation. Usually compared to business strategies for overall effectiveness of the operation.

If we go back to the seven steps of the Manufacturing System and the hierarchical concept of linked computers (see Figure 4.1), we can see how the common database serves the needs of the organization to provide the information required to run the factory. Figure 4.2 shows the eight databases (Table 4.1) superimposed on Figure 4.1, illustrating this point.

There are certainly other important databases that could be added to the common database, and many businesses do so. Consider, for example, the personnel database or the payroll database. Both are important for conducting business and both certainly have data that can be related to the "two knows." The personnel database usually contains skills levels of individuals. This is certainly pertinent to the scheduling algorithms necessary to determine cycle times. Likewise, the payroll database is necessary to obtain accurate cost evaluations. This illustrates what is meant by stating that there really is not an all-inclusive generic example of "the" common database, but we can define bare minimums for most cases.

We have now examined external and internal database architecture, hierarchical computer systems, and some generic database contents for manufacturing; but, we must keep in mind the reason for employing CIM philosophy is to improve profitability. We improve profitability by becoming more productive. As managers, we are always interested in productivity enhancement. It is the way to stay competitive and enjoy a successful business tenure. Let us examine how CIM allows us to control our operations

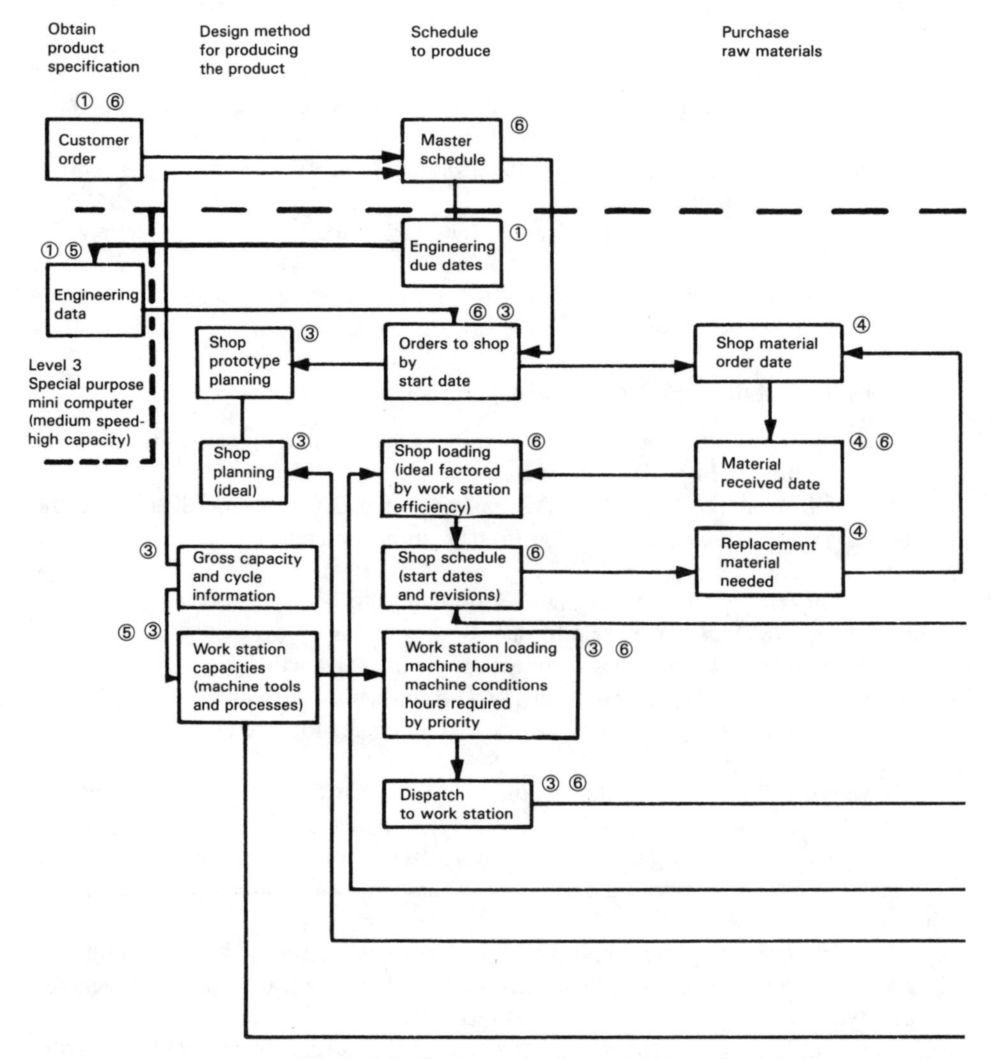

Figure 4.2 Common database and hierarchy linked computer flow

better, that is, how we benefit from employing synergistic uses of computer hardware and software.

It must be stressed once more that the one and only reason for CIM is to improve profitability by means of productivity improvement. That means we want to lower operating costs. Our purpose here is to show how CIM lowers operating costs.

Recall that in chapter 2 the concept of the seven steps of the Manufacturing System was presented. It was stated that all companies providing a service or making a product must comply with it. Those who understand the seven steps and make an effort to optimize them are more effective and, hence, more profitable. Those firms that deny

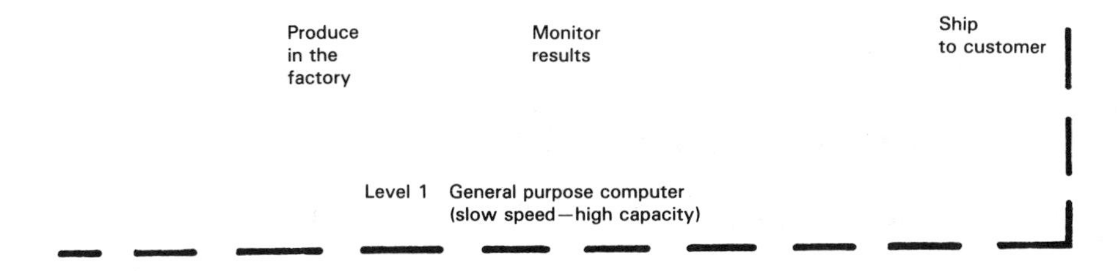

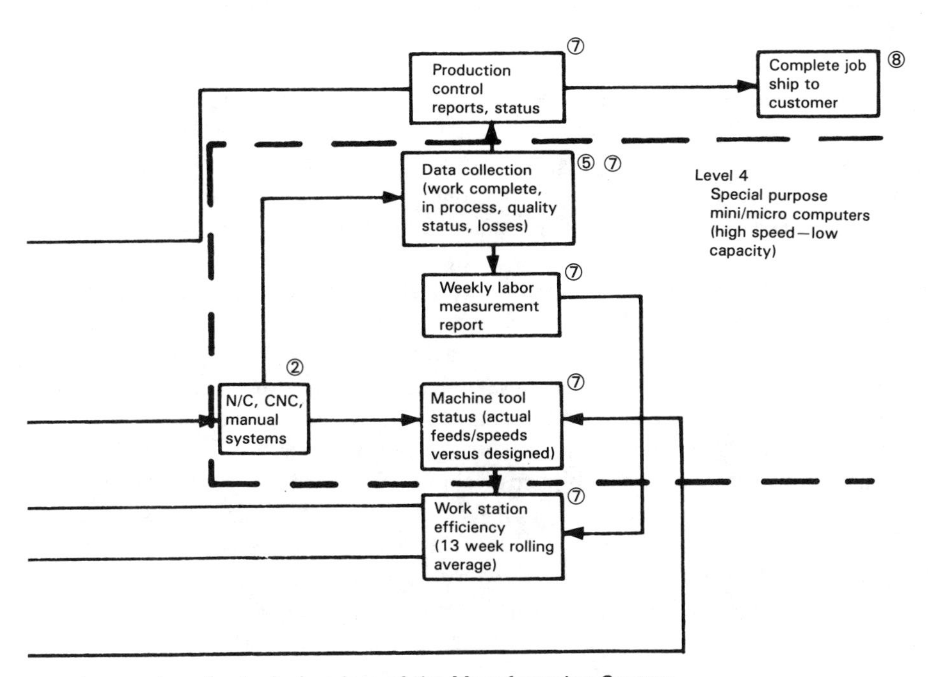

superimposed on the typical variant of the Manufacturing System.

the logic of the seven steps or, even worse, fail to understand that the logic exists at all, do worse, and many fail. So the answer of how CIM lowers operating costs lies in how companies understand the seven steps and how they automate them, link them, and eliminate waste by doing so. To illustrate and support this argument that proper uses of databases, architecture, and structure of computer components leads to optimum control of business operations, let us look at the information coordination made possible by automating the seven steps. Let us see how this improves our ability to manage and thus achieve greater profits.

A typical job shop (chosen for the illustration because job shops represent 70% of all North American manufacturing) operates in the following manner.

1. An order is received.

2. It is put into the manufacturing master schedule by the marketing department and agreed to by the manufacturing department.

3. The engineering department produces a tailored set of specifications for processing and fabrication.

4. The manufacturing department takes these specifications and makes up task instructions for various work stations.

5. Concurrently, the manufacturing department orders materials.

6. The specific sub-components are scheduled for the factory.

7. Work is done and verified.

8. Finally, the completed product is shipped to the customer.

9. While this is going on internally, the marketing department (and to some extent, the engineering department) is communicating status and answering customer questions.

The common factor through all this activity is information, information as to what to do and when and with what equipment and how precisely and when to stage the raw material so as not to be too early or too late and how to react to nonplanned changes requested by the customer and how to fix errant work while maintaining schedules. It is a veritable beehive of information given and requested and reacted to. The better we can do in transmitting information and keeping it current, the less likely we will have foul-ups and the more likely the factory will work at optimum effectiveness. To enhance the probability of optimum effectiveness, recall chapters 2 and 3. The seven steps of the Manufacturing System are able to work in a dynamic mode made possible by proper selection of internal and external database architecture, the structure of the computer hardware throughout the factory, and the information stored in the common databases; that is, the desire is for precise efficient management of information. With the ability to manage information at a much higher level of competence, we minimize the time it takes to send and receive confirmation of receipt of the data, thus making it possible to carry out an operation faster. This reduces the cycle time necessary to do the job. Since time is really money in the business world, reducing cycle time reduces costs and, hence, improves profitability. With CIM, achieving "excellence in communication" is much more likely, which means there is a much greater chance that the job can be done faster and more accurately. The linking of the seven steps through the architecture, structure, and common databases is CIM and is the reason we can provide the product at a more rapid and accurate rate. This is productivity improvement. What follows is a more specific example of CIM philosophy in action:

Concerning common databases again, for example, many functions have a stake in how the component parts they manufacture are to work. Design engineers are to be able to do the scientific analysis of the geometric shapes, to build in proper safety margins. Manufacturing engineers are to design the manufacturing process so it is compatible with the design functionality and to make sure tolerances are met. Here we have examples of two functions using the same information but in a different manner. If we have one unified database for this information, then we have optimized our probability for success. This is only possible if we have a common database that is maintained by the primary function (in this case design engineering) but accessible to other functions with a need to know. The goal of the Manufacturing System has always been to do this. But, forced to work in a static environment, this was not very likely. CIM allows a dynamic environment that fosters excellence in communication. This naturally leads to greater accuracy and thus to improved productivity and lower costs.

Computers make group technology practical for managing data, which in turn makes CIM more effective. If we are going to have efficient communication, then we need a methodology for cataloging data. Group technology requires computers to manipulate the codes on which it is based. But, once the computer is applied to group technology (GT), it does wonders for the concept of CIM, making CIM a much more practical entity. Group technology will be explained in a later chapter. It is so necessary for effective CIM that, while GT is not a true subset of CIM, it is important to understand GT for its inherent benefits traditionally associated with modern CIM-conceived enterprises. Our purpose here will be to show how a business based on CIM principles is even more effective if GT is employed. We will link GT to databases.

We improve productivity by means of CIM through the coupling of common databases with GT. GT can be defined as a method of systematically searching for sameness in manufacturing techniques. This makes larger production runs possible and commonality of tooling achievable, allows dedication of machine tools to families of similar parts, and efficient utilization of many designs and plannings. Common databases include large quantities of design, marketing, manufacturing, and financial data pertinent to all contracts worked by the company. After a relatively short period we find that specific aspects of new jobs are very similar to aspects of previous jobs for which data are stored. It is evident that if data previously stored and previously used are similar in nature to new data and new work requirements, then redundancy may exist. If we can eliminate this redundancy, then we will have a much more efficient business operation.

By using GT classification and coding schemes as an index for the common databases, recall of the data on the basis of design characteristics, manufacturing characteristics, and other specific features that may be of particular use to the company is allowed. Their recall process optimizes CIM. For example, when asked to design a new component, design engineering can use the GT code to quickly search the design database for exact matches (components designed before for other customers) or to a design similar enough to require only modifications to make it useful for the current requirement. Finding a similar design shortens the design cycle; getting an exact match eliminates it. This shortened design cycle means more work can be done in a specific time period. Shortened design cycle means work can be produced faster with the same resources, which yields higher productivity for lower costs. The benefits derived for design engineering are also available for manufacturing engineering in the creation of methods and plans. Similar benefits also are achievable for purchasing and quality control. The coupling of GT principles with databases, as we can see, is a great enhancement to our ability to optimize the control of business operations. We are maximizing our ability to manage through optimizing excellence in communication.

Let us examine another example of the capability to optimize business operations through CIM. Once more, enhanced capability is created by using communications excellence made possible through software architecture, hardware networking, and use of common databases. This example will be the productivity/quality measurements and evaluations capability offered by the CIM philosophy.

The seven steps are a logical progression of activities leading from customer order to filling that requirement and delivering the product. From a managerial perspective, the most important item is as precise control of the process as is possible. Control allows decisions to be made at appropriate times to minimize costs and maximize output. It is gaining control and maintaining it for which management strives. Throughout the history of business, management has continuously struggled to gain and maintain control, sometimes successfully, often not.

Before computers, monitoring and controlling the seven steps were done by manual means. Control centers were set up to track progress much like military command posts. Information as to progress according to a schedule was requested by means of the best communications tools available. A nineteenth century army relied on dispatch riders to the various battlefield sites. In the precomputer era, managers used production followers and expeditors to keep track of specific order progress. The results in both cases were slow and of variable accuracy. The history of human armed conflict is full of miscalculations leading to defeat or missed opportunities. Learned men and women have devoted entire careers to trying to understand why generals and messengers behaved the way they did. If the history of business had been written with the same vigor as the military counterpart, we would probably find the same human failures. Why did manager X react to the information of component Y being 6.2 hours behind schedule as he did? Should he not have known that his decision or lack of one would cost the company the entire profit margin for the job? We will never know these answers, but we do know that manager X was plagued with untimely and late information that resulted in wrong decisions being made all too often. Manager X knew this but had no choice but to console himself with the fact that he probably made enough right decisions to, in balance, rate a satisfactory performance for his company. The problem for manager X and his military counterpart was information befuddlement, with lots of it coming in with infinitely variable accuracy and timeliness. How to sort through this and use the best information is the question.

One way precomputer era managers attacked the information quandary was through the use of "tally sheets." The tally sheet method requires all who do value-added work or material movement and storage work, as well as those who check for the accuracy of the work performed, to record what they do and send the records to the control center. This creates a history of what has happened and, it is hoped, those in charge have the opportunity to digest and react to the information in a timely manner. A whole host are employed to gather the tally sheets and react to them. Traditionally, we call these people production control specialists, expeditors, or job followers. Their function is to keep on top of all jobs assigned to them and to keep them on schedule. This amounts to manual data collection and responses in accordance with a preplanned schedule. If the volume of different types of activities each follower is assigned to track is small, the system seems to work. The problems arise when complexity and numbers of jobs increase, as you would expect. There is also a consistency problem. Each production control specialist comes to the job with a different level of competency and drive. Therefore, the entire production performance, the amalgam of all the production control specialists driving their product lines, goes forward in fits and starts, in a very uneven manner for those trying to control a business. In the recent past things have gotten even worse with the introduction of numerical control (N/C) machines. These machines, run by independent non-networked microcomputers, have sped up the process of making parts to an alarming rate such that expeditors are having serious problems keeping up on the tally sheet front. N/C machines predate the CIM philosophy and, when used as originally conceived, create "islands of automation." N/C machines properly linked with computer-controlled information systems are a true revolutionary development in our ability to make parts less expensively at a higher quality level (this will be discussed in depth in later chapters).

The information and control problem is one that is solved by CIM. With proper networking of information made possible by means of software architecture, linking of computer hardware equipment, and the use of information such as schedule and comparison-to-schedule stored in common databases, it is possible to choreograph a campaign to optimize use of resources and materials to produce the products. The computer, drawing information from the common databases, can create coordinated schedules. By using hierarchical-arranged computers throughout the factory, expeditors can enter data concerning the status of projected completion and compliance to design specification. Another computer, at a higher hierarchical level, then can access the common databases in which the data have been deposited and compare actual accomplishments to plan. The resulting evaluations can be presented to management for action decisions. How does this compare with the precomputer era? Obviously quite favorably. We are optimizing collection of data, analysis of that data, and the ability to make correct decision (see Figure 4.3).

What was just described in the previous paragraph is the genesis of what is now called manufacturing requirements planning (MRP), a subset of CIM. MRP will be

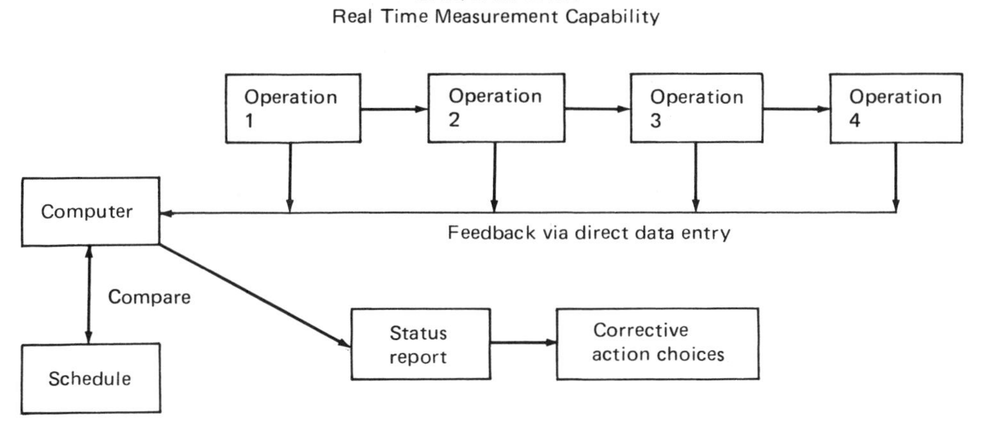

Figure 4.3 Real-time measurement capability of CIM. (From Daniel T. Koenig, *Manufacturing Engineering: Principles for Optimization,* Hemisphere, Washington, D.C., 1987.)

explained and debated in later chapters. It is important to note that what was just described leads to the capability to manage in a dynamic real-time manner. With the manual precomputer system, unless it was relatively small, the emphasis was to maintain schedule at all costs. The schedule was the plan. Because of the high variability of information available, it was much more prudent to stay with a known plan than to try to change and inform all those involved as to exactly what the change was and of any new actions required. With a CIM database, this is no longer true. Change can be radiated accurately by means of the common database, making it possible to be very flexible in reacting to business conditions. This breakthrough results in true dynamic response capability versus manual status. Dynamic response, flexibility, and adaptability are all means of describing what happens when the CIM philosophy is applied to control of business operations. Simply stated, they get better and the profit picture improves.

This chapter concludes with what I call setting the stage for management understanding of CIM. We have looked at the structure of CIM. We see how the interweaving of software and hardware creates pathways for information to flow. We relate the uses of common data to solving the manufacturing control needs. We begin to understand the benefits to management that can be achieved by means of implementation of the CIM philosophy. The rest of this book is devoted to explaining the contents of the CAD/CAM triad, how CIM is implemented, and related topics that fit CIM into an overall managerial perspective.

Chapter 5

The Place for Numerical Control and Factory Automation in a CIM System

This chapter is devoted to the machine-process control triad or CIM. From the viewpoint of the public, this represents what CIM is. As we have discussed so far, we know it is only an important component of CIM but it is by no means the entire entity. We will demonstrate the importance of numerical control (N/C) for modern manufacturing concerns, its basic technical foundations, and how it relates to the other two legs of the CIM system triad. When an N/C system evolves along the premise of the CIM philosophy, it becomes a flexible manufacturing system (FMS). FMS requires its own versions of controls, again evolved out of the CIM philosophy. We will investigate these from the management viewpoint. FMS demands companion automation and control of materials sent to and from the various manufacturing cells. This too becomes a topic for this chapter. Finally, FMS and automated material handling systems connote a degree of total automation, so this discussion will conclude with factory automation and the concepts of the factory of the future.

Only 5% of job shop activity and 30–35% of flow manufacturing activity directly involves machine-process control. This is the value-added portion of manufacturing. The rest of the time is nonvalue-added and does nothing but add to the cost of the final product. One of the primary goals of CIM is to grow the value-added activity percentages by eliminating communications, storage, and throughput wastes in the other two legs of the CIM triad, then optimize the variety and throughput capabilities of the value-added portion (the machine-process control leg).

Machine-process control is the visible portion of CIM. N/C machines, automated transfer devices, and robots on the factory floor are technologies to marvel about.

However, while this is important, we already know that it is insufficient by itself as a total CIM solution. Therefore, it is important as managers that we be aware that the design and implementation of the equipment need to be totally compatible with the needs of the other two legs of the CIM triad. I state this right at the beginning of the discussion of machine-process control because it is easy to come under the spell of these wonderful technologies and lose track of for what they were developed in the first place. It is desirable to use N/C and all its derivatives; but, given the choice of N/C and manual control systems or computer-based control systems and manual machines, we must always opt for the latter.

Machine-process control consists of the hardware devices used to create the product, transport, and store partially finished and finished product, display the proper tools to the work stations, and the associated computers and software necessary to make these devices work. The range of devices and software is enormous, from basic N/C machines to robots through flexible automation. It is not the intention of this book to produce engineers qualified to design and build these devices, but to present an overview of the pertinent technologies so that we can manage their implementation and future development directions. We want to be sure that companies we represent get optimum value for capital invested in these technologies. For that reason some time will be spent describing the fundamental principles underlying the theory leading to N/C and its descendants.

The introduction of the computer to control a machine tool or a process is the most profound change in industry since the invention of the steam engine. Just as the steam engine released industry from the confines of water power, which limited the size, shape, and site of factories, the computer liberated manufacturing from the vagaries of skilled artisans. The N/C machine allows every operator to perform at the level of the best master machinist. It does this by capturing the best methods of operation into its computer memory by means of programming. The computer allows many machines to perform exactly alike since the directions given the machines can be identical. This means complicated parts can be replicated less expensively and, hence, superior products are available to a larger number of consumers. Taken to its logical conclusion, this means the standard of living of the population may improve. This is indeed a profound change brought about by introduction of a new technology. Let us delve a little deeper into machine tools and what is significant about N/C.

A machine tool can do one or more of the following operations:

- Drilling and boring: Making holes of small to large diameters.
- Turning: Making externally round pieces.
- Milling: Making flat pieces with a rotary tool.
- Planing, shaping, and broaching: Making flat pieces or concave pieces with a stationary tool.
- Grinding: Shaping either curved or flat surfaces with an abrasive.
- Forming: Deformation of materials into desirable shapes by bending, pulling, shearing, and pounding.

These operations are the practical universe of what can be done with what is commonly referred to as engineering materials (steel, wood, plastics, ceramics, aluminum, copper, titanium, and so on), and they make up the totality of our modern society. The one thing all the machine tool subsets have in common is the requirement to be directed through an intelligent control. That intelligence until fairly recently has been human, at least in a direct cause-and-effect mode. With N/C, the directing mode becomes the computer, and we do this because we can count on the computer to do a better job of conveying instructions to the machine than the average person can. This is true because the instruction given to the computer is more likely to be superior than the best operators are likely to be available to run the machine. One set of instructions derived by one master machinist can be duplicated to run a multitude of N/C machines. Contrast this with the fact that that same master machinist can only run one machine at a time. This gives an idea of the scope of the revolution we as managers must cope with for the benefit of our firms.

The secret of this revolution is the method of instructing N/C machines to do our bidding. It is called parts programming. To make the N/C machine work, it has to be instructed in a manner that the computer understands. There are many choices. We can instruct it by means of machine language (e.g., converting our desires into a binary code directly), or we can use some sort of intermediary form. The former would require a computer systems engineer. The latter is the work output of the computer systems engineer and an industrial engineer into a user-friendly software package that allows the user to use plain language to instruct the computer on how to run the machine tool. This is parts programming. It tells the machine how we want it to work on the engineering material. With this instruction in its memory, the computer generates a sequence of electrical signals to be sent to the machine's control unit (MCU), which in turn activates the gears, motors, and transducers necessary to create the appropriate machining motion in contact with the work piece. The machine can rapidly step through the various commands to do the desired work, ending up with a finished product usually faster than the skilled human can (because the N/C machine does not have to stop at intermediary points to check for accuracy the way an operator would).

The method of creating a modern parts program is relatively standard. An industrial engineer studies a work piece drawing, looking for a logical sequence of operations. This is no change from the pre-N/C era. For example, the engineer would determine the type and number of cuts (tool passes over the work piece) and the types of tools to be used. For each cut in the pre-N/C era, the depth of infeed into the material is put on an instruction sheet for the operator. With an N/C machine this information is transferred into points in space on a cartesian coordinate system with a datum point being known on the machine tool itself. These points in space represent a tool path for the computer to follow (e.g., have the cutting tool transverse across the part to be machined). Listing of all these points in space creates a manuscript for the N/C machine to follow. Descriptively, this tool path is shown in Figure 5.1.

Creating an N/C program is an engineering function. However, to be practical, we want non-engineers whenever possible to make N/C programs for specific uses. This is analogous to supervisors and operators deciding how to make desired parts on manual machines. To do this, engineers have created what is called "family of parts programming." These contain all the generic information required to make a common shape on a type of N/C machine (a lathe, saw, vertical boring mill). Suppose we wanted to turn a crankshaft on a lathe. We would calculate all the lengths, diameters, fillet radii, and other features as points in space, then factor in tool offsets and infeed parameters, and in essence create a unique manuscript program to be fed into the MCU. For a complex piece like a crankshaft, this could take quite a while, even for a qualified engineer. Instead, we could use a generic program that will accept the specific parameters that make the crankshaft unique, much like using popular spread-sheet programs such as Lotus 1,2,3 to create specific control charts instead of designing the chart ourselves and writing a program to enter it into a computer. For the crankshaft we want to find a program that is compatible with the N/C lathe and at the same time is compatible with the shape we want to produce. There are literally hundreds of such programs available. Most require basic shape and other geometric parameters to be entered, usually where there are discontinuities in shape direction, such as where the diameter changes or holes are encountered. This makes creating a manuscript for a specific part really quite simple. No longer do we need an engineer to do this, but, even more beneficial, we do not even require a programmer. The operator, the supervisor, or even the manager's secretary can enter the data.

Family of parts programming makes N/C very practical and very useful. In some cases it is even more versatile than manual machines because no one really has to have the skills necessary to run them, at least not at the level of the skilled machinist. It is a simple matter of having the correct family of parts program, relatively simple instructions on how to use the program and run the machine tool, and the ability to load raw material onto the machine and off-load finished parts. This may seem too good to be true, and that is probably correct. There are always complications, but, really, they are of much less complexity than learning the skills necessary to run a manual machine. The capability now is available for virtually any firm to make extremely complex parts without the full-time support of expert engineers and programmers. Think of what that means to the competition equation worldwide! From the managerial viewpoint, having N/C equipment requires a whole new game plan. It offers tremendous competitive flexibility and opportunity if used correctly. But before discussing the managerial impact of N/C, let us discuss the simplicity of making parts by means of N/C with a family of parts program.

Figure 5.2 represents a typical input sheet for a family of parts program used to create a manuscript for an N/C lathe. The product would be a shaft of some type with up to five diameters, five lengths, and four fillet radii. The number of combinations of dimensions allowed by such a program is enormous, well over a quarter of a million. This indicates the creative ability the factory has to make rotors, shafts, or whatever. The designer is certainly not constrained by the equipment. The process of creating the manuscript is simplicity in itself. The person desiring to make the rotor has only to assure himself or herself that the number of physical dimensions does not exceed the number of inputs allowed, in this case, five diameters, five lengths, and five fillet radii. Then, record these values from left to right on the input sheet, and type the values into the computer. The computer (usually a personal computer or equivalent) creates a program in a language the MCU of the lathe can understand. The program is fed into the N/C lathe by means of a tape, floppy disk, cassette, or any other means desired, and the machine is now ready to make the part. How simple, how powerful a competitive tool. We have utilized an expert programmer and an expert engineer to design the

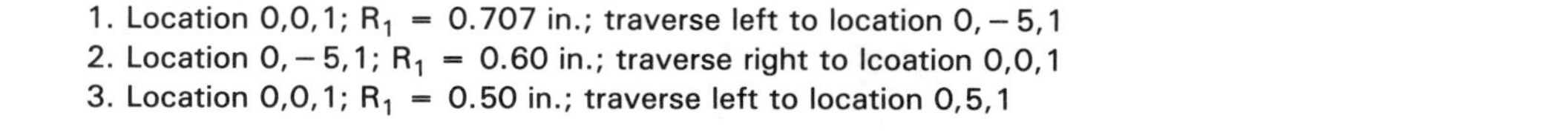

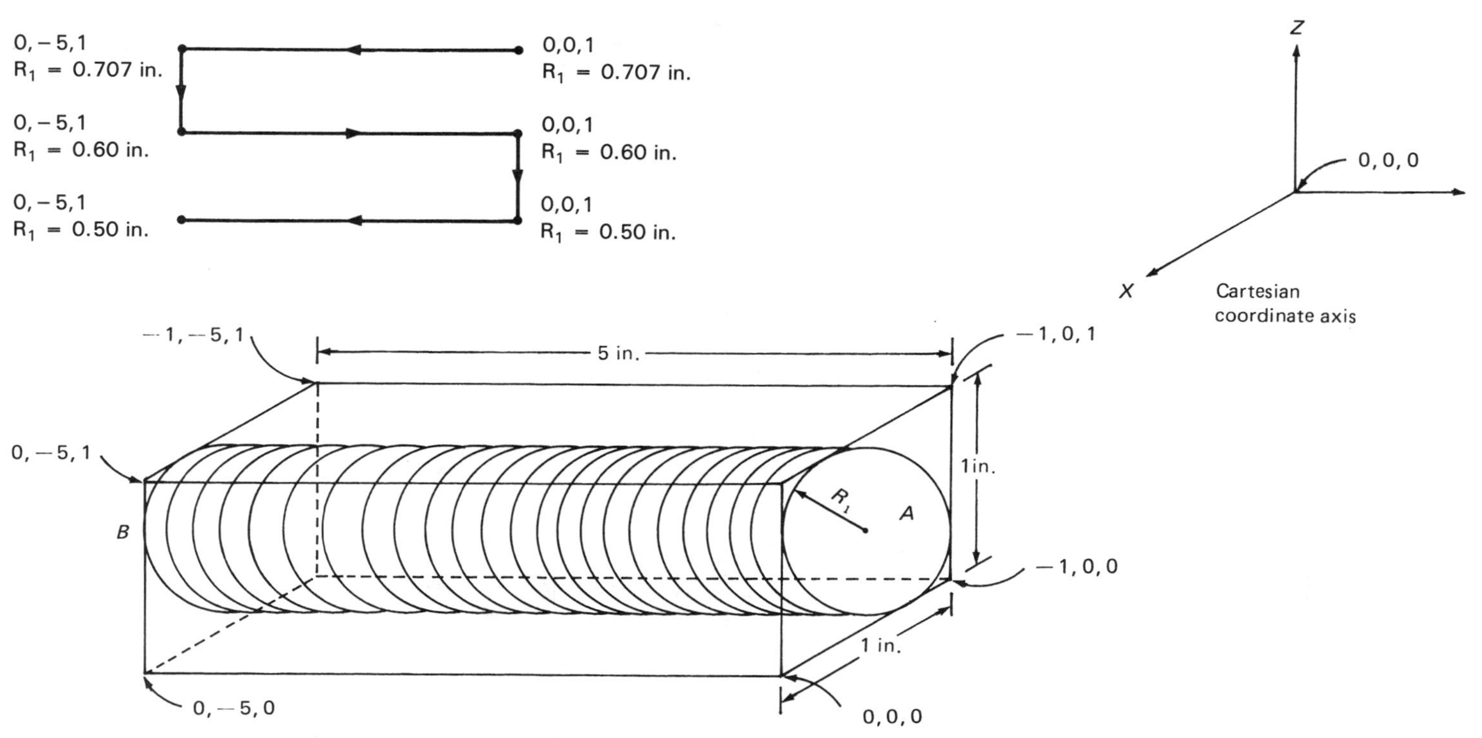

Figure 5.1 Square-faced block converted on a lathe to a round dowel using lathe tool path. (From Daniel T. Koenig, *Manufacturing Engineering: Principles for Optimization*, Hemisphere, Washington, D.C., 1987.)

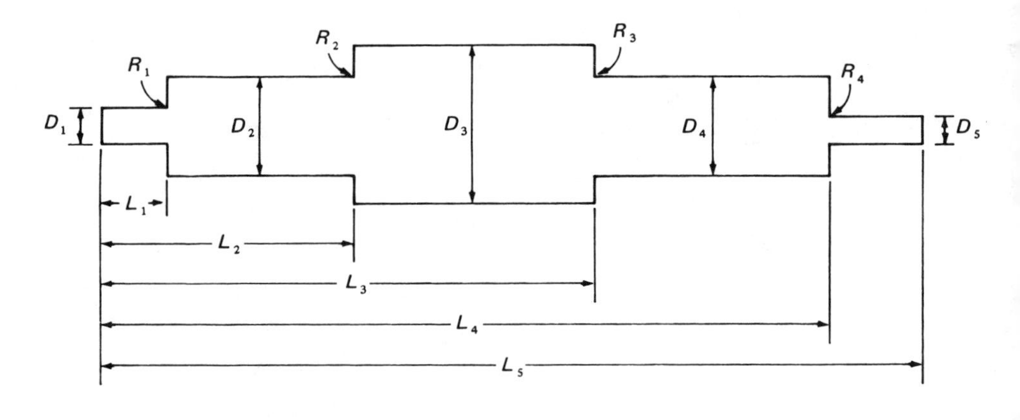

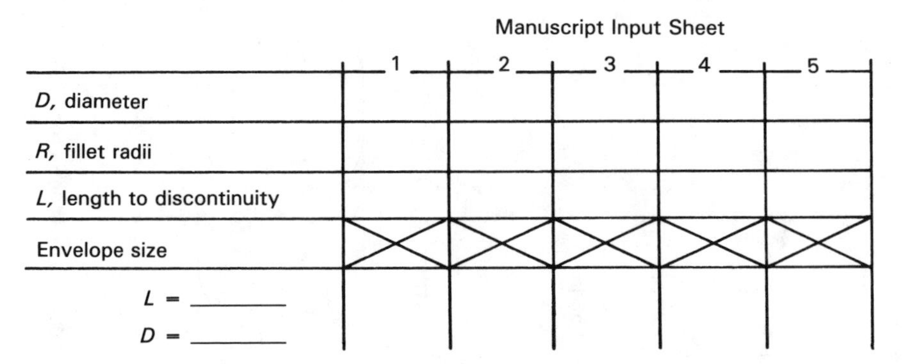

Manuscript Input Sheet

	1	2	3	4	5
D, diameter					
R, fillet radii					
L, length to discontinuity					
Envelope size					
L = ______ *D* = ______					

Figure 5.2 Generalized outline for producing a rotor shaft using family of parts program. (Adapted from Daniel T. Koenig, *Manufacturing Engineering: Principles for Optimization,* Hemisphere, Washington, D.C., 1987.)

process to make a unique part, and those people may not even be on our payroll. We can compete with virtually anyone with minimal investment and literally turn out huge varieties of one-of-a-kind parts at a fraction of the cost of a manual operation.

So now, what is the difference between a manual machine and an N/C machine? There is only one relevant difference, that is, how the machine is controlled. A manual machine is controlled solely by the operator taking his or her direction from drawings and supplemental methods instructions. The N/C machine is controlled by the MCU, which is a miniature computer, taking direction from an input program. The secret of the success of N/C over manual is the ease of creating input programs. This leads to distinct differences in the management approach to using N/C machines. In addition to the obvious integration of the equipment into a CIM solution (e.g., a synergistic whole), there are certain considerations managers must keep in mind for proper use of this equipment.

• N/C has the capability to produce at a rate two to three times that of similar manual equipment.

• N/C maintenance downtime owing to inherently more complex control devices is approximately 90% higher than for manual machines.

• Types of factory skills change with introduction of N/C. More emphasis is placed on methods personnel with programming understanding than on skilled artisan-type machine operators.

These considerations lead to a different slant in managing when N/C equipment is available. Since there are very few factories available with all N/C machines, and there may never be (an example would be hard automation where superefficient transfer lines with dedicated equipment are in place to make very large production runs at minimal costs, such as in soap manufacturing), managing the use of N/C will be discussed in comparison to managing the traditional manual mode.

Establishing planning guidelines as to what parts are to be made by means of N/C equipment is necessary. It is not advisable to allow the selection to be an on-the-spot or spontaneous one. Always plan to load the most efficient machines first, whatever type they are. These become the base-loaded machines dedicated to making the core of the factory's products. Usually, the N/C machines tend to be the most efficient because of their superior throughput rates.

As part of the guidelines, review parts to make sure that making them on an N/C machine gives the company an advantage. The advantage usually comes about if the part is complex or if large quantities are needed in short periods. The differentiation of complexity or quantity for the choices is and probably will remain a subjective evaluation. If there is an either-or decision to make, typically when we can have either large volume quickly or complex parts, the decision usually favors technical considerations of making highly accurate complex parts.

The next consideration is how to enhance the use of N/C to achieve its inherent advantages. This comes about by means of structuring the work in preparation for manufacturing and also by understanding the subtle and not so subtle changes in roles.

As with any desire to make anything in a factory, the first step is to determine machining methods and tooling requirements. This is done by a planner. Once accomplished, it is necessary to make a conscious decision as to whether or not to employ N/C equipment in the manufacturing process. If the answer is yes, we divert from the normal process and use the services of a programmer. Here the planner-turned-programmer translates drawing and process instructions into the input format required by the family of parts program. Now, what is different here? We have assumed that the planner is capable of doing the family of parts program inputs. This means management has made the commitment to train personnel in this new skill. In addition, management has consciously (we hope not unconsciously) made the decision to dedicate manufacturing engineering resources to the task of pursuing the application of more family of parts programs to more parts. This is necessary so that the N/C machines become more useful and offer more choices for applications to management. By this method of training and searching for more uses, the N/C equipment gradually becomes the preferred equipment.

One more point should be made here. The training and familiarization does not stop with the planner and the engineer. It continues right on to the operators. They no longer have to be skilled machinists, but they most likely have to operate more than one machine at a time. So, it is important that operators be familiar with the capabilities of the N/C equipment to the point that they become error detectors and know when the machines are malfunctioning. They should be trained in corrective action initiation and some trouble-shooting investigation. Operators become technicians rather than craftsmen in the N/C age.

No discussion of numerical control should go beyond this point without introducing robots. Robots, for all their hype, both commercially and in literature (including fiction), are N/C machines. They are specialized N/C machines but not necessarily particularly advanced in the concept of what an N/C machine is supposed to do. Machines are designed to transform engineering materials into useful products. Some do intermediate steps only. Some do intermediate steps and finishing steps as well. The one thing our machines do not do well is put parts together. We do have assembly machines associated with hard automation, but these devices are capable of performing only one or a very few operations automatically, and the rest are done manually at the end of the line by an operator. So, robots are an attempt to use computers coupled with transfer-type machines to do assembly, coating, and transporting processes normally performed by operators. Robots are an expression of the desire to mimic the motions capable of being performed by humans so we can replace humans in dull and dangerous tasks. Let us look at where we are right now in the development of this evolutionary cousin of the N/C machine.

The concept of control becomes more complex for the robot than for the N/C machine because it has more degrees of freedom. The N/C machine, though capable of relating to three-dimensional space through the cartesian coordinate system zero datum point, is constrained by the physical realities of the machine. It has guide rails, jacks, threaded worm drives, and other physical devices that limit the motion of the cutting tool. This is necessary, of course, to make parts of very precise tolerances. We use the moment of inertia of large massive structures to ensure positional accuracy. Robots tend not to have point-in-space crutches. They literally have no wall to hold onto nor to confirm their location. This means the device needs a navigational system to tell itself where it is and also a motion-inducing and -stopping system to hold it where it wants to be, in three-dimensional space, when it gets there. These are serious engineering problems that, as yet, have not been totally solved.

If we are going to replace humans in doing assembly work, then there is a definite need for accuracy. Approximation is not good enough. Imagine the idiocy of saying it is sufficient to drive a nail into a wall beam that is within 6 inches of the connecting joist. There would be lots of nails driven, but very few making the connection. I do not know if it would be a good idea to live in a house built like that.

Unfortunately, the controls found in most N/C machines are literally no more accurate than that. There is no need to be more accurate because of the physical constraints, which we use as guides, that inherently exist because it is a machine tool. Robots do not have this advantage, so the need for more precise control systems is a necessary and specific requirement. This has led to experimentation with a television-based vision system, infrared sensing, and other imaging systems that can be related to known boundary conditions for allowing the robot manipulator arm to know where it is in relationship to where it should be. To date, location control has not been solved well enough to allow the robot to work more precisely than the human assembler, painter, stacker, or whatever.

In addition to the location problem, robots must deal with the desired degree of finesse for the application. For paint spraying robots, this is no problem. Once the robot knows where it is, the paint is applied. But, if we are interested in assembling components, a highly sensitive degree of feel is required. Otherwise, the power of the manipulator may literally shear bolts and strip threads. A human has a remarkable range of force and dexterity that can be applied to the assembly process. Couple this with simple mechanical advantage tools, and it is tough to compete against a man or woman doing assembly work. This means robots must be given a tactile feel through a force-resistance measurement feedback system or some other comparable system. While in theory this is possible, in practice it is difficult to accomplish. Even if it is accomplished, we have to ask, is it economically justifiable to want to do it? The answer, of course, depends on the task. For space exploration, absolutely yes; for putting together motors, it is questionable. This means that for unpleasant or dangerous situations, the robot is suitable. Knowing factories, this means robots are not likely to be as prevalent and dominant as their cousins, the N/C machines, at least robots as we know them today.

There are many undesirable manufacturing jobs that people are required to do to produce desirable products. For this reason we see a desire to make robots work at levels approaching human efficiency so men and women need not do uncomfortable, dirty, or dangerous work. The problem is how to overcome the current primitiveness of robots to accomplish these tasks. The answer lies in fixturing and designing the task to be more forgiving of tolerance errors. Many applications for assembly where people overcome sloppy placement of devices have had to have more precise location devices for accepting the part so a robot could be employed. This is commonly called "crutching the robot," but it is necessary so that a robot of a cost level that is practical can be used to eliminate the use of people from undesirable tasks. This is the current status of robots. They are used most frequently to replace people in undesirable jobs. Robots generally do not work faster or more accurately than do people, and, hence, they are not yet creating a competitive advantage.

The overall objective of CIM is to use computers to optimize the seven steps of the Manufacturing System. We have seen that with N/C, and, to some extent, with robots, managers have been able to control the value-added portion of manufacturing more closely. If we were to do no more than this, then we would be losers. We have not satisfied the desire to optimize the seven steps. Instead, we would have created islands of automation, no matter how spectacular they may be. We would still be grossly inefficient in feeding those work stations and in removing material from them. Thus, we must consider how we optimize and perhaps automate the support mechanisms (the other two legs of the CAD/CAM triad) to make them as efficient as N/C. To do this, we have to explore the border area between N/C and the other two legs.

To explain this border area, it is necessary to introduce the terms "flexible automation" and "flexible manufacturing systems." They are current buzzwords and, as such, really have no precise definitions. I suppose that is to the benefit of computer hardware and software companies as well as machine tool builders to repackage their offerings, perhaps with a few new wrinkles, to sell their products. These two terms denote, and I chose to describe them as being, synergistic N/C, in that they are N/C machines capable of multi-process functions that are somehow tied to external demand sequencing logic. So, flexible automation and flexible manufacturing systems are N/C machines that are somehow tied in with scheduling, data collection, planning, and material handling systems. Therefore, for simplicity, we can say that flexible automation and flexible manufacturing systems are the merger of MRP II concepts with N/C to get greater synergism in performing the seven steps of the Manufacturing System.

Is this good? I think it basically is. We have inputs into the common database describing tool paths derived from the geometry defined by the design engineer for N/C machines to follow. In concert with this, we have scheduling algorithms in the same common database that are set by the master schedule but can react to actual performance as reported by the data collection feedback system. This, in turn, triggers or coordinates the release of orders to the various N/C machines by calling for material to be delivered on an as-needed basis. Finally, it initiates the work on the N/C machine. It turns it on. Flexible automation appears to be able to think, that is, starting and stopping machines based on data feedback. Obviously, it is not thinking in a classic sense, just responding to electronic stimuli. If everything is working right, and if we were clever enough to program in all the possible contingencies, plus having a common database large enough and fast enough, then this should work. We should have an automated but flexible factory. In theory, it is fine, but, in practice, the complications cause significant downtime and frustrations. So, what it all means is that the concept of synergistic N/C is laudable and theoretically feasible, but we are not there yet.

Some of the reasons we are not really in the flexible automation era yet are as follows:

- *Hierarchical control of computers in the factory:* As discussed in chapter 4, there are many different needs for computer sizes, memory storage configurations, and speeds as dictated by specific activities carried out at the work stations. It is very difficult to define those activities absolutely. Thus, to compensate, we need to have higher capacity computers than are really necessary for most of the tasks. This means that costs can skyrocket, making the entire project uneconomical. For hierarchical control to be successful, we need low-cost flexible minicomputers and microcomputers.
- *Communication difficulties between very different work activity centers:* Imagine engineers, planners, expediters, managers, cost analysts, quality assurance specialists, and others all trying to dictate their needs to each other on a real-time basis and all using the imprecise human languages. A few misunderstandings are going to occur. An automated factory cannot tolerate even a small level of misunderstanding. Fortunately, much work is being done to alleviate this problem. Communication standards and protocols are being developed. Led by General Motors and Boeing Corp., MAP/TOP (manufacturing automation protocol/technical office protocol) is being developed to standardize how we will communicate between work activity centers. The fact that they are on mod 3.0 and introducing 4.0 establishes the difficulty in achieving univer-

sally workable standards. In the area of language, we are fast learning how to replace imprecise human language with the use of group technology codes, where the meanings are precise enough for our purposes.

- *The need for ad hoc query capability in the databases:* As was discussed in earlier chapters, relational-type databases are required as we depart from very rigid applications of computers into the field of reacting to situations that are not preconceived (i.e., solutions preprogrammed). As was discussed previously, relational-type databases are not yet mature, in that they are slower than is desirable and still costly. Fortunately, the computer industry, seeing the profit potential for efficient relational-type databases, is spending significant capital in solving these problems. Computers designed to work efficiently and quickly with relational databases are on the horizon, and software improvements also are creating more efficient relational databases.
- *Lack of trained personnel to manage flexible automation factories:* This is the what came first, the chicken or the egg dilemma. Managers cannot spend large sums of money training people for future systems if they are not likely to get a reasonable return for their effort. Since viable flexible manufacturing systems do not now exist, there is little desire to train people in the theory and use. If people are not trained, then the implication of what does exist does not grow very quickly. For example, there are sufficient multi-purpose N/C machining centers linked with their own computer-controlled material handling systems available on the market today. There are perhaps a thousand or so in use as reported by popular trade journals. But this is very small penetration of the potential market, especially after a decade of development. The major reason for slow expansion is the lack of trained personnel to manage these systems, a classic Catch-22 situation. This is not flexible automation, but a step up the evolutionary ladder, and it is having a difficult time reaching a critical mass. The solution will probably come from leadership supplied by the technical universities and the professional societies as they reach out to the political and economic leadership constituencies.

If we are to have flexible automated factories in the future, combining the N/C capabilities of the machine tool with the automated storage, dispatch, and off-loading of the material to be worked on is necessary. To do otherwise would be to perpetuate the island-of-automation syndrome. It does us no good to ride along a dirt road and then rapidly advance on a superhighway only to be bottlenecked again onto a dirt road at the end of the superhighway. That is exactly what happens when we have an N/C machine outstripping its supply line. If the N/C machine is a machining center, then the supply problem becomes geometrically worse. Since machining centers combine more than one material forming/removing process at the same work station, they are prodigious users of material, and keeping them stocked is a significant task. For this reason, coordinating the material needs of the work station with the inventory control and the dispatching activity is very important if we are to have the success (in the nature of productivity improvement) that the machining center is capable of achieving. To achieve this success, a specialized application of computers to transporting and storage systems has come into being. This is known as "automated material handling systems."

For classification purposes, we list it under the machine/process control leg of the CAD/CAM triad. However, for control purposes, automated material handling is closely linked with MRP II, which is a major component of the production and measurements control leg.

Automated material handling can be divided into two parts:

1. Automated storage and retrieval systems (ASRS)
2. Automated guided vehicles (AGV)

ASRS connotes computer-controlled warehouses, and this is exactly what they are. They are buildings or large modules within buildings that contain individually identified shelf space linked by transport systems to and from a major input-output dock (shipping and receiving station). Material is received at the dock, cataloged by the computer as either outgoing or incoming, and added or subtracted from the inventory log. The space where it is supposed to be stored is assigned, or, if the material is being dispatched, that space is then determined to be vacant and available for the next compatible set of incoming material. The material is moved about in the ASRS by computer-controlled cranes and transporters such that the modern ASRS has very few people, if any, within the recesses of the structure.

Automated guided vehicles (AGV) are driverless delivery systems from stocking points to work stations and back again. Sometimes they travel a series route until finished product is delivered to the stocking point. Traditionally, we think of AGVs as little robots delivering parts, mail, or whatever to desired locations. They are definitely computer-controlled, but only a portion can be considered robots.

For flexible manufacturing systems, we desire AGVs to be loaded at the ASRS and then to deliver material in a flexible path to any one of a number of machining centers. There are ASRSs that automatically load AGVs that then independently go to their assigned destinations. Most AGVs are not automatically loaded. People load them and then instruct the computer that the AGV is loaded. The computer then selects the route, knowing what the load consists of, to the destination. Most often, AGVs are wire controlled. The device follows a track wire buried in or pasted onto the floor, sort of like a railroad. The computer energizes the required route, and the AGV travels to the destination. Pasting the track wire to the floor is favored because it allows for future re-routes as the nature of the factory changes.

The true robot AGVs are not too numerous. They differ from the guide wire variety by their ability to determine where they are in two-dimensional space (none that I know of fly yet). This is usually done by a beacon signal or a light that is within line of sight of the AGV in the factory. Knowing the geography of the factory, the onboard microcomputer can triangulate the location. These AGVs usually also have feelers that keep it from having serious collisions with people, equipment, and other obstacles it will encounter in its journeys. Typically, these feelers are accelerometers that measure resistive force. If the force is at or above the threshold level, the computer program instructs the AGV to stop, back up, change direction by a pre-determined amount, and then try again. This is a relatively mature technology and is rapidly gaining acceptance in factories.

Regardless of type, once the AGV arrives at its destination, it is necessary to make the material transfer. Some are done manually and some are done automatically as part of the N/C program. The choice depends on the degree of automation required and the economic justifications. AGVs moving to and from work stations appear to indicate true flexible manufacturing systems in action. They normally are not such indicators. Seldom are they dispatched as a result of the data collection system showing the need for material. This is because the MRP II systems are usually not fully integrated with the automated material handling systems. Typically, the output of the MRP II system will be a schedule dispatch list, which is independently acted on by the personnel responsible for material handling. This human interface is usually required for reasons stated previously about why flexible automation is not a reality yet. However, this monitoring by the MRP II system and the dispatching of information for someone to take action is vastly superior to the manual systems of the past.

This now leads us to consider the future. We have looked at fundamental N/C, its current state of evolution into flexible manufacturing systems (FMS), the nature of robots as to how they fit into the N/C universe, and the auxiliaries to feed the work stations, namely automated storage and retrieval systems (ASRS) and automated guided vehicles (AGV). Where is all this leading to? As managers we must consider what the trends mean and where the research is most likely to lead manufacturing industry. Beyond a doubt, concepts of CIM will prevail because they do lead to a better ability to optimize the seven steps of the Manufacturing System. The concepts provide for a competitive advantage. The question is, what is the concept of the factory of the future, and when will change be evident?

I believe the factory of the future will need to be more automated but very flexible. As true skills such as master machining and woodcraft decline, the desire for products produced by these skills does not necessarily also decline. So, the demand has to be fulfilled. The only way to do this is through use of the ascending skills (i.e., an N/C programmer) in making computer-controlled machines "sing," to produce a beautiful thing through the application of mathematics driving cutting tools through points in three-dimensional space. This means there will be economic forces providing incentives to solve the technical problems described previously in this chapter. The questions for managers, then, are how fast and how extensively can this be done?

The noticeable advent of flexible automated factories will probably occur faster than we are prepared to handle, or, at least, the technology will be mature enough to allow implementation in an economic manner by aggressive competitors. Looking at Figure 5.3 suggests that implementation of N/C and all its descendants has been a geometric progression. Manual to N/C dominance took 25 years. N/C to CNC took about 5 years to mature. CNC to FMS is happening right now and is essentially mature to the point of feasible economic implementation. The last step is FMS to truly automated factories. How long will that take?

For managers to understand the magnitude of what is going on, it is helpful to define terms as a progression from N/C.

- N/C: Stand-alone machine controlled by a built-in microcomputer called a machine control unit (MCU).

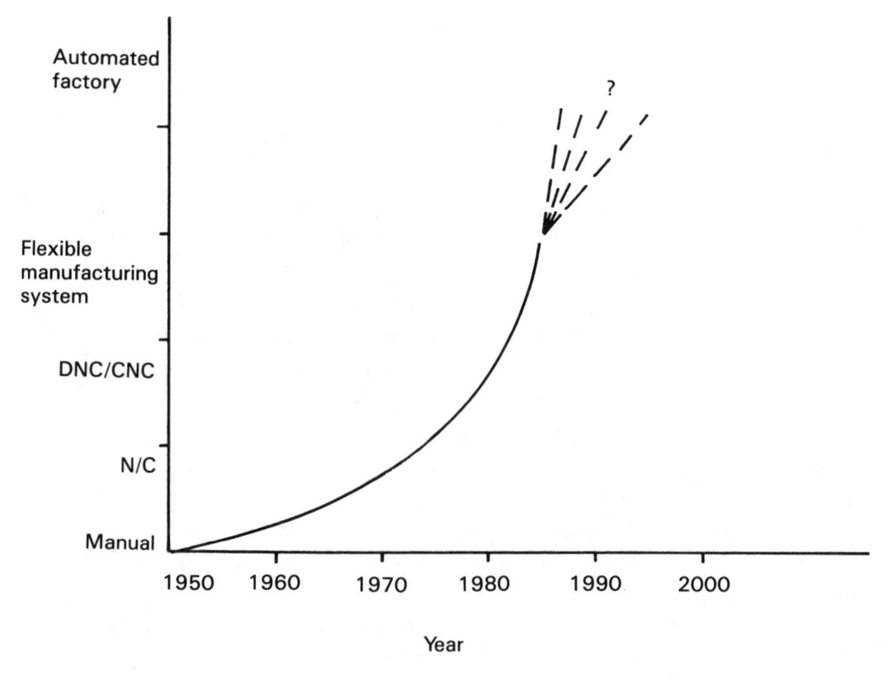

Figure 5.3 Progression to the automated factory.

- CNC: Grouped N/C machines; usually for similar parts manufacturing; coordinated by computers.
- FMS: Grouped N/C machines; usually multi-turreted tool changer types, with coordinated material handling devices; closely coordinated (semi-controlled) by computers.
- Automated factory: CIM, automated linkage of the seven steps of the manufacturing systems implementation; grouped FMS systems, controlled by hierarchical computers, with decisions based on the total inputs of the CAD/CAM triad (design and planning control, production and measurements control, machine/process control).

It is evident that the technology builds on itself and that the interrelationships between the legs of the triad become more important. This is a salient point that managers cannot disregard.

From this definition, we can see that the automated factory is, in essence, grouped FMSs (machine/process control) linked with design and planning control and production and measurements control. FMS is the ultimate in machine/process control. It gives the manufacturing organization the ability to optimize small lot sizes because of its super flexibility, thus approaching the efficiency of flow manufacturing. This flexibility comes about because of the multi-tool characteristic that makes the machine capable of performing combinations of basic machining functions (recall the earlier portion of this chapter defining the basic machining functions). This gives the planner the capability of mixing and matching operations on the machine without losing value-added time if different setups have to be performed. Because of this, FMSs are economical for medium size production runs of 200–20,000 parts per year. This represents well over 50% of all jobs produced by manufacturers worldwide. I believe it is easy to extrapolate that if we have the other two triad legs tied into the control of the FMS, then economic lot sizes of one become feasible and that efficiencies of flow manufacturing are approachable.

The FMS provides manufacturers with the capability of making whole ranges of families of parts (group technology definition of parts sharing similar geometric and/or manufacturing characteristics) because it is very easy to program the machine to produce the parts without setup changes. It is the difficulty of setup that separates flow manufacturing from job shop manufacturing. So, minimizing the difficulty of setups in job shop environments with FMS allows the development of parity between the two and gives an advantage to job shops through its greater flexibility.

FMS allows for easier scheduling algorithms because many operations are possible at the same work station. This minimizes routing and movement considerations, which, in turn, minimizes the number of computer inputs and outputs to the scheduling and data collection systems. Here we have a simplification caused by employing a more complex machine, an interesting irony. The irony is compounded even more when we realize that most benefits derived from this ultimate expression of the machine/process control technical development is in the other two legs of the triad. The benefits lie primarily in the reduction of the 95% non-value-added portion of the manufacturing cycle.

Table 5.1 shows in summary form where benefit can be calculated to support capital requests to implement FMS.

In chapters 3 and 4 we saw how the hierarchical control of the manufacturing system was desirable for optimum results. The use of FMS demonstrates this. In fact, FMS would be much more complex than need be without the hierarchical control concept. If we think of FMS specifically and the automated factory as a series of interconnected FMSs, then we can see that Figure 4.2 can be thought of as basically the design of the product, the purchase of materials, one or more FMSs, and shipping

Table 5.1 Benefits of FMS

	CAD/CAM Triad		
Benefits	**Machine/ process control**	**Production and measurement control**	**Design and planning control**
Operate equipment around the clock	√		
Minimize direct labor		√	
Minimize lead time			√
Reduction of in process inventory		√	
Reduce tools and fixture requirements		√	
Obtain high flexibility		√	

the product to customers, where all steps of the manufacturing system are factored into the hierarchy control. In fact, numerous experts see FMS and, thus, a good deal of the automated factory, being controlled in a three-tier hierarchy of computers: top, administrative control; middle, supervisory control; bottom, machine monitoring and control.

If we relate this to the seven steps of the Manufacturing System, we see that all but two are covered.

- Administrative control: step 1, obtain product specification; step 3, schedule to produce.
- Supervisory control: step 2, design a method of producing the product.
- Machine monitoring and control: step 5, produce in the factory; step 6, monitor results.

The only steps not done internally (or at least partially internally) to the FMS control then are step 4, which is purchase raw material in accordance with the schedule, and step 7, which is ship the completed product to the customer. Step 1 makes the basic assumption that design and data as a result of that design are separate and different. This would mean that design and manufacturing engineering must remain independent. This may not be optimum and must be looked at carefully.

In summary, then, we are looking at a progressively automated factory coming into being, one that employs computers in a hierarchical way to control the CAD/CAM triad, linking the seven steps of the Manufacturing System in an optimum manner, and that will be as automated as deemed desirable. It is management's task to take advantage of these developments to enhance their organization's performance or possibly to ensure its survival.

Chapter 6

The Use of Computer-Aided Process Planning in a CIM System

Computer-aided process planning (CAPP) is the automation of the methods, standards, and planning function of manufacturing engineering. By its general nature, instructions as to how a task is to be accomplished, we can expect that CAPP will be a prodigious user of computer memory. Thus, true CAPP has had to wait for the development of faster, less expensive, larger capacity computers and highly manipulative databases such as the relational types to be practical. This development has arrived, and bottom-up CAPP, going from incremental building blocks of data to completed cycle times for entire assemblies, is now a reality. This chapter will trace the concepts of planning from the theoretical base to application as a major component of the CIM system. CAPP is a portion of the design and planning control leg of the CAD/CAM triad, and we will see how it relates to the traditional seven steps of the Manufacturing System, other components of this leg, and the other two legs of the triad.

One of the main functions of manufacturing engineering is to produce a manufacturing plan to make the company's products. Success in this function is measured by how close to optimum the factory is running when compared with the system the engineers have designed. Here we are making the assumption that the constraints faced by the engineers are that they have to optimize the use of "existing" facilities. This means there is a very definite premium in understanding the task at hand and in devising a way to operate that is as efficient as possible. So, planning is not a by-guess-by-golly procedure. It is a discipline that calls for utilization of engineering principles to discover the one "true" way of making the product.

In enterprises that create value-added goods and services, there are two absolute prerequisites for success. I call these the "two knows": (1) know how to make your product, and (2) know how much time it should take to make it.

These are the bedrock of success in the manufacturing and the service industries. Knowing how to make your product or deliver your service means having an expert understanding of how the product or service comes into being. This means understanding the underlying principles of science engaged in the design and the matching principles necessary to construct the product or service so it will function as contemplated when completed. With this basic information, it is possible to structure correct methods for cycling raw materials through various processes to completion of manufacture.

The first "know" deals with science and process selection. The second "know" defines the size of the factory. A basic premise of manufacturing, in accordance with the seven steps of the Manufacturing System, is scheduling. To schedule, it is necessary to know discrete operation cycle times and order of operations. Knowing this, it is possible to schedule, or load, the various work locations (commonly known as work stations or centers) within the factory and its vendors' facilities. The "two knows," therefore, provide the basic information for all strategies of managing the output of products from a factory. This is process planning, that is, merging the science of making the product with the required sequence and cycle time. The technique that engineers use to do process planning and related activities is an adaption of the scientific method.

The scientific method is a procedure that virtually all science-based disciplines use to try to discover the truth. It is a means of evaluating observations by means of establishing a hypothesis and then testing that hypothesis to determine if the observation is explicable in accordance with what is held to be a scientific truth. If the testing of the hypothesis does not yield the expected results, then the hypothesis is modified on the basis of the latest observations (collected data). Eventually, if this process is repeated enough, the hypothesis can be declared correct and the observations then said to be understood.

This process has been very successfully applied to manufacturing engineering when creating optimum planning for producing products in a factory.

Table 6.1 illustrates this special adaptation of the scientific method to manufacturing engineering. Note that this is because of step 2 and the recipient portion of step 1 of the seven steps of the Manufacturing System. We are doing the detailed work of designing a method of producing the product in a manner that will yield the optimum way of making it. In a CAPP system, the computer is used to augment the manual approach of optimizing by means of the scientific method. It makes use of group technology code to minimize redundancy, and it is dynamic in that methods can be changed with changes of the design by means of the common database. CAPP is by all means a member of the CIM database concept and is a module of the CIM system.

Before delving into details of CAPP and the various interrelationships, it is necessary to understand why a scientific approach to process planning is necessary. As alluded to in the previous chapter concerning the development of N/C, the work force throughout the industrial world is changing, changing as our outlook toward factory work is changing. People throughout history have striven to make life more amenable

Table 6.1 The Scientific Method Applied to Manufacturing Engineering

Traditional scientific method	In terminology applicable to Manufacturing Engineering
1. Make observations	1. Study the design
2. Develop a hypothesis	2. Create the method and a written plan
3. Test the hypothesis	3. Try the method at the work place to produce prototypes
4. Make revisions to the hypothesis based on the test	4. Analyze the prototype production method for lessons learned and potential improvements
5. Test the revised hypothesis	5. Try the revised method at the work place to produce another series of prototypes
6. Reach a workable conclusion	6. Iterate method improvements until the percent gain reaches a pragmatic limit; then finalize the method for production

and less physically stressful. This is particularly true in our social relationships concerning how we provide food, clothing, and shelter for ourselves and dependants. We are looking for easier ways of earning a living. This means there are less of us who are willing to put up with the discomforts of working in factories and, hence, the advent of N/C and the trend toward the automated factory. Unfortunately, the skills needed to run the machines and processes are also eroding. We no longer have the luxury of using expertly skilled people in the desired quantities in factories to produce products at the quantity levels desired. This erosion has been going on for several decades. To keep our factories going, we have had to do a better job in defining the steps necessary to create the products. This is the task of the manufacturing engineer, and it becomes more complex as the basic factory skills decline. This erosion process and the inherent increase in detail required to achieve status quo is called the "guild hall to industrial engineering–based factory process," which means methodology needs to be more precise, which in turn requires better and more detailed planning. It means optimizing the use of the scientific method as applied to manufacturing. This means the only practical solution is CAPP.

Figure 6.1 shows the change in skills experienced by typical industries in the past few decades as social outlook toward skilled factory jobs has changed. This presents the challenge and the opportunity for CAPP.

Now let us see how CAPP has evolved over the decades as a means of adapting to the social changes occurring in industry and briefly review the history of computers in manufacturing industry.

The first industrial application of computers was scientific computing translated into design engineering use. Design engineers have desires to solve complex equations quickly and accurately. As engineers know, developing a model of a physical entity to understand how it works or should work can result in very complex equations, especially if some rate of change of a physical quantity with respect to other constraints is anticipated. The results are often expressed in equations that are difficult and at best tedious to solve. Since computers can do calculations extremely quickly, and through ingenious programming difficult equations can be solved, the computer becomes a

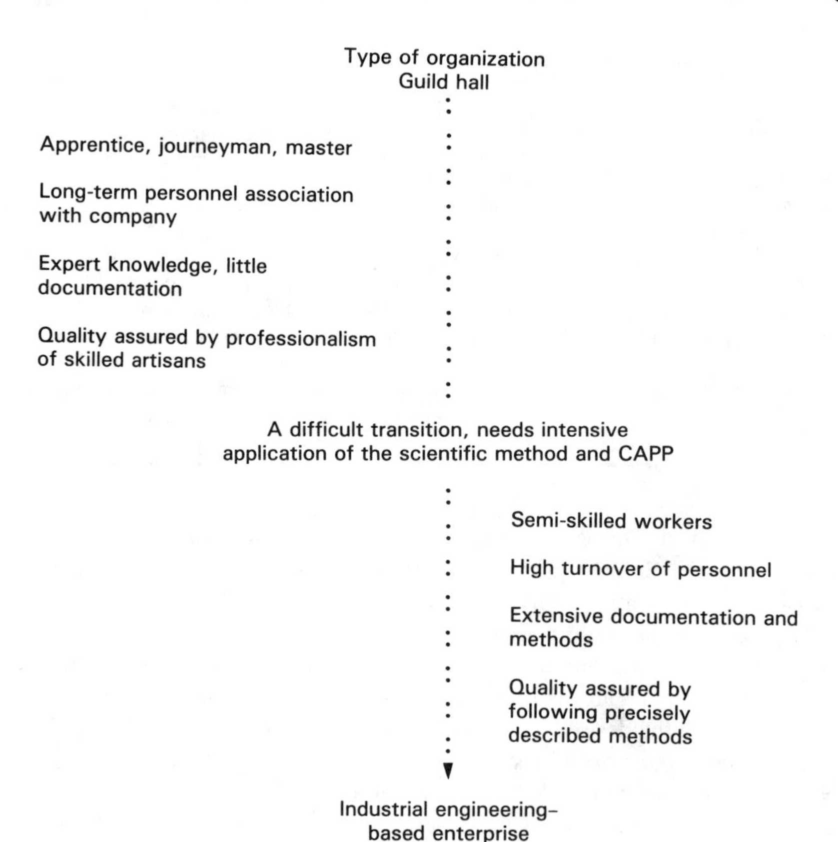

Figure 6.1 Guild hall to industrial–engineering based factory process.

powerful tool for the design engineer. So computers were introduced into industry by means of design engineers borrowing from their scientist cousins.

The second application of computers in industry was in finance. Here, the desire was to do the monotonous task of double-entry bookkeeping much quicker and at a higher degree of accuracy. The computer's ability to do simple repetitive arithmetic and rarely make mistakes was a natural match of need to solution availability.

The third application of computers in industry was the development of numerical control (N/C). N/C is an exercise in three-dimensional geometry. The definition of points in space and then deciding the differences between two locations is nothing more than simple arithmetic. To make machine control practical, these arithmetic operations had to be done very quickly and, hence, the need for the computer. Manufacturing engineers either used the same geometric data as design engineers in setting up their models for stress and strain evaluations and solutions, or input the data independently. The computer then did the arithmetic required to instruct the machine how far it should go in any direction and in the sequences desired.

The one thing these three applications had in common was that they were not integrated with each other in the least. In the early 1970s, these were three distinct and independent usages of computers in factories. Then, natural curiosity and a need to communicate led to shared data and shared data use experiences. With N/C, the need for geometry paralleled a need of designers and, hence, the shared database was developed. Interactive graphics developers showed how creating a design on a video display tube (VDT) in essence is an exercise in three-dimensional geometry. Since N/C requires three-dimensional geometry as an input, why not create a tool path program at the same time the designer is creating a program to draw the electronic image on paper? This is what happened, and we had the beginning of the CIM common database.

From the financial view, manufacturing observed lots of simple data being manipulated quickly for general ledger and cost accounting purposes. It quickly became evident that scheduling is a very similar process. Instead of adding dollars, the scheduler is working with quantities of parts and cycle time, and the computer makes it much easier to handle lots of data and constantly changing data. This is the genesis of the whole idea of manufacturing resources planning (MRP). It also becomes evident that finance needs much of the same quantity of cycle time information to calculate costs and, it is hoped, profits. So the concept of shared database was further spread.

The next piece of the CIM concept of shared data was a result of another manufacturing need, the desire to have very rapid (dynamic) input of instructions and to receive just as rapid and accurate feedback on the results of implementing those instructions. To accomplish this, manufacturing borrowed some techniques from finance and developed some new ones. The status of parts completion is as vitally important to finance as it is to manufacturing. To calculate ongoing costs (period costs), finance needs to know the status of completions in the factory. This information is requested of manufacturing periodically. It is in essence schedule updating. Finance only needs it periodically, while manufacturing could make good use of it continuously. Since manufacturing was already providing data on a periodic basis, why not find a way to do it continuously and only give finance the information when they requested it? The continuous data had sufficient value to manufacturing in validating schedule plans with actual achievements and allowing timely corrective action that it became apparent that development was justifiable. This led to user-friendly data collection terminals and to database design that allowed easy access and ad hoc query for problem solving. This data residing in the common database has enormous potential use (as has been demonstrated) for both manufacturing and finance.

CAPP, which is vital for any scheduling algorithm to work, is the only totally original computer system developed by manufacturing. CAPP is based on the concept that each activity required to do the defined work can be mathematically described as a series of forces and moments related to velocities and accelerations of humans and machines. Therefore, we can accurately describe the true time it should take to do any process. We can then string these times together and come up with a cycle time that can be used for scheduling purposes. This concept existed prior to CAPP but was too tedious to do for any but the most important analyses. The process, developed by the pioneers of industrial engineering, is commonly called "principle of motion economy" and is the basis of scientific time standards. Manually, this is an extremely difficult chore to do. It requires defining all the motions involved in doing a job and determining the motion paths and then the time required to do each individual motion. In practice, it involves looking up values for incremental motions, adding them up, and coming up with a portal-to-portal time to accomplish the task. This is then put into the plan for the work to be done at the work station. This is practical if only one or two different items are made at that work station. Any more and the complexities mean it would take too long to come up with the time value. For this reason the use of scientific time standards, while known to be of significant value, is not universally employed. We get by in the majority of applications with estimates instead. However, with the computer's ability to do many calculations rapidly, engineers have rethought this technique, and we now have the capability to accurately relate work sequence times to planning. We can effectively achieve the second "know" as easily as we can discern the first "know."

We will describe this CAPP process in greater detail in its modern, mature, version. For those interested in learning more about the evolution of CAPP, I recommend reviewing my book, *Manufacturing Engineering: Principles for Optimization* (Hemisphere, 1987). I believe we now have a basic view of these major components of the CIM database and are prepared for the explanations of how CAPP functions.

Planning is the development of strategy to achieve an aim. To be complete, planning consists of two components: a direct line approach and a set of contingency plans to overcome unforeseen occurrences. When we plan anything, from a picnic to a space launch, we normally include some form of contingency. The contingency for the picnic may be as simple as an alternative date or location to contend with inclement weather, while the contingency plans for the space mission may be many multiples of complexity beyond the primary or direct component. The amount of contingency depends on the degree of risk the planner is willing to take against successful conclusion of the task. The risk of failing to have a successful picnic is negligible, and, hence, the degree of complexity of the contingency planning is also negligible. At the other end of the spectrum, a failed space launch is a risk no one is willing to take, or, at least, the risk must be minimized to a very low level. Hence, the contingency component of planning will be very complex and well developed. Manufacturing task planning falls somewhere between these two extremes.

In manufacturing, the goal is to have both the primary and the contingency planning capable of producing the desired results. The primary plan will consist of the best way possible to achieve the results. This usually means employing the best facilities available in the best sequence. The contingencies, by default, consist of using less efficient facilities or less optimum sequences.

Sometimes, there is more than one contingency available. If this is the case, the contingencies will be ranked with the highest being the closest to the primary and the lowest being the farthest away, in terms of efficiency. In all cases, all versions of the planning are capable of making the product in accordance with required physical specifications. All versions of the planning, however, may not result in meeting the required cost levels. The closer to the primary a contingency is, the more likely it will meet the financial goals. A major reason contingency planning is now practical beyond one or two alternatives is the computer. It would be impractical to have many contingencies for a manual system. The ability to find, brief personnel on, and execute obscure or lower level contingencies would be poor at best. With a computer system, the planning dispatches, primary plans, and all the contingencies are resident in the common database of the CAPP system. So, we have the same ability for the computer to execute a primary plan as, say, contingency number sixteen.

CAPP belongs in the design and planning control leg of the CAD/CAM triad. Our analogy of the triad being a three-legged stool, where each leg is necessary to fully support the stool, correctly implies that CAPP needs to interrelate with the other two branches of the triad. CAPP provides basic inputs to manufacturing resources planning (MRP II), a major component of the production and measurement control leg. CAPP also receives information back from that leg as well as from the machine/process control leg. We are looking at integrated systems, and while we are discussing CAPP, we have to keep in mind always that it does not operate in a vacuum. In fact, acknowledging the existence of the other two legs will help us to understand better the nature of CAPP.

When we think of CAPP we envision a set of sequenced instructions dispatched by a computer to work stations involved in making a part in a factory. This is insufficient. To be complete and therefore useful, the instructions must also contain optimum cycle times. Why does a process planning system need cycle time information? Is it not sufficient to just sequence operations for MRP II and then use experienced operation times for scheduling? The answer is no. To optimize the manufacturing procedure, to know what the factory is capable of producing, it is vital that the process plan be defined in terms of optimum times to do the specific required process tasks. This gives management a foundation on which all business strategies can be built. It tells us exactly what we can do with the existing facilities. It gives us capacity information. This engineered time standards concept, based on the principles of motion economy, is the first step in process planning. It is the building of a library of accurate times for performing the discrete processes used in the particular factory. These are the basic building blocks from which a total method to produce a product is developed.

The second step of process planning is to define the method of making the product and then create instructions for the factory to follow. Once the method is choreographed between human and machine, it is merged with the library of engineered time standards. The results are called operations planning. We will discuss this concept of merging standards with sequenced instructions later under the overall subject of bottom-up planning.

What is described in the two preceding paragraphs is exercising the "two knows." We have defined the process for developing the descriptions of how to make the product and also how long it will take to do it. Now let us see how process planning integrates with other functions and how CAPP allows all this to happen in a dynamic mode. By using the philosophy of the "two knows" we create a bottom-up process plan. This is the optimal way of producing parts in a specific factory. It should be obvious that in a manual mode it would take considerable effort to plan for an entire factory, especially if that factory is required to make a large variety of nonstandard parts. For this reason, we must use computer assistance if we want to operate effectively in a dynamic mode. CAPP is designed to do just that. CAPP links various databases to output work station process planning. Figure 6.2 shows the relationship of these databases in a generic CAPP system.

The input of the CAPP system is customer requirements from the marketing

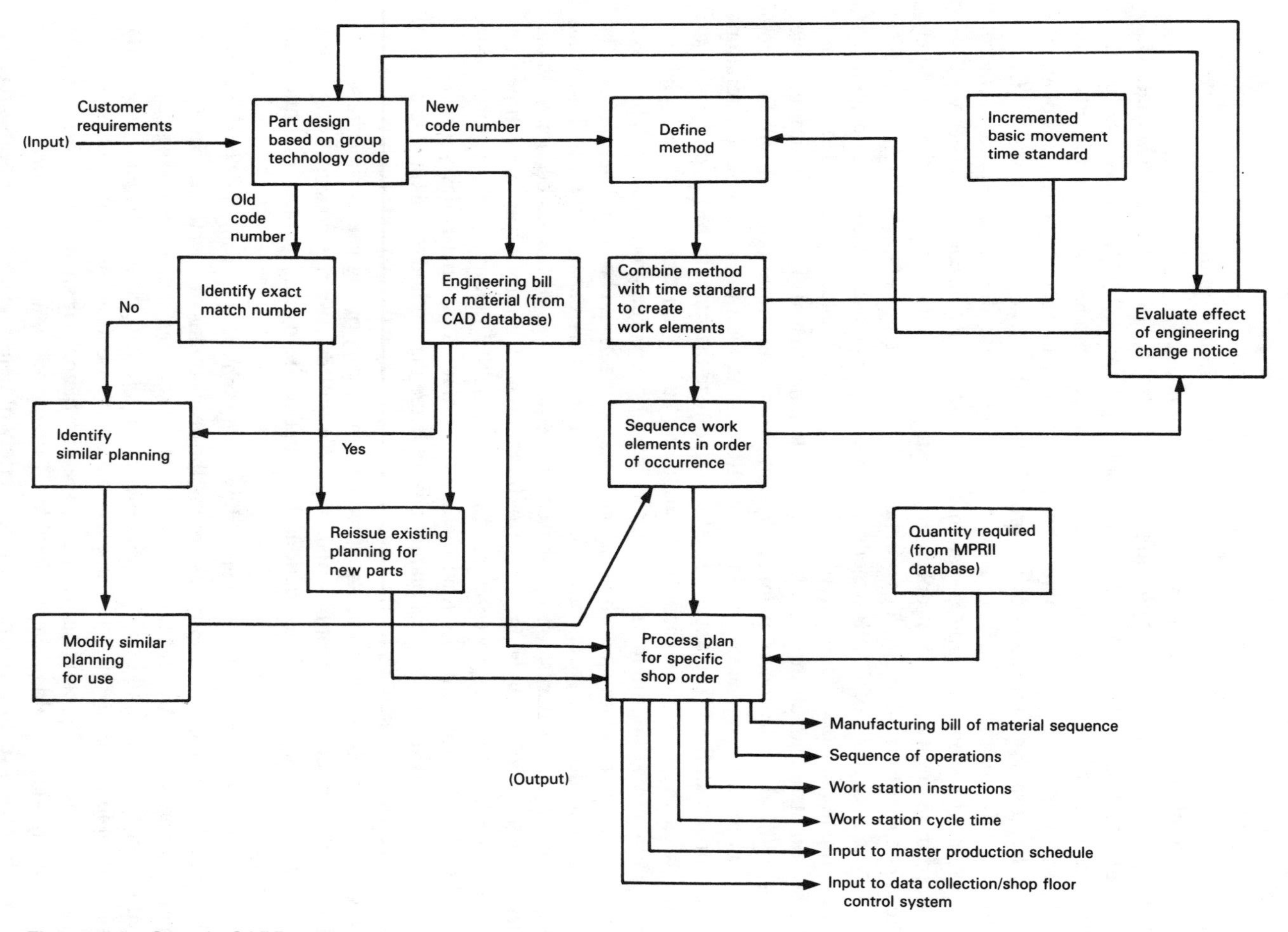

Figure 6.2 Generic CAPP system.

database. Therefore, a true CAPP system is not an internal manufacturing database. As we can see from Figure 6.2, there are several input sources outside of manufacturing. The design engineering database is an important contribution and, through group technology, can set the focus for factory instructions. Since we need to work in a dynamic world, it is important for manufacturing to be integrated with engineering. For example, note in Figure 6.2 how the evaluation of the effects of engineering change notices are fed from parts design to methods evaluation and back to parts design. This is an interactive loop demonstrating the integrated relationship between engineering and manufacturing. This type of iteration would be impossible without computer-driven databases because of the vast quantities of data to be evaluated on such short time scales. CAPP is a dynamic system that allows for this optimal approach available only through integrated databases.

To elaborate further on the value of the integrated dynamic approach afforded by CAPP, let us discuss the benefits made practical by CAPP.

Improve response time to customer needs and/or inquiries by more efficient factory planning and scheduling. Through knowing the optimum cycle time based on engineered standards it is practical to schedule the factory with less reserved time for unknowns and, thus, obtain faster turnaround. For inquiries, engineered time standards–based planning creates known values for quotations instead of estimates.

Improve overall product quality through better documentation and control of the manufacturing process. The application of the "two knows" requires greater attention to detail; therefore, possibilities for making errors are brought to light much sooner. Hence, preventive and corrective actions can be taken sooner and usually at a lower cost.

Improved integration with engineering by means of group technology and best method information. The ability to link the effects of potential engineering changes on manufacturing processes prior to actual dispatch of the task to the work station creates significant improvement in overall effectiveness. The trial-and-error technique of manufacturing is greatly reduced and, hence, profitability is improved. This attention to detail at the design stage saves more than tenfold corrective costs later on.

Improved customer quotation validity through more accurate cycle time, personnel, and material analysis. This should reduce the relative dollar value of quotations, thus winning more bids. With more accurate data, contingency factors can be made significantly smaller and, hence, lower cost quotations can be provided to customers.

Reduction of support personnel through improved communications between functions. With integrated databases required by true CAPP, there is less of a need for logistics personnel to keep information flowing between functions. We no longer need an army of clerks receiving, recording, verifying, and retransmitting information to various subfunctions. This can now be handled in a dynamic automated database mode.

Bottom-up CAPP, which we shall see is total CAPP, is the methods/time standards/planning portion of CIM. All of these items are communications items. Methods communicate the technique for doing work. Time standards delineate the calculated time for doing a segment of work. Planning combines the technique and the time to accomplish the work into sequences and sets of tasks. All of these items are instructional in nature. The more precisely we do each, the better the final results. If we can link method to time standard to planning, then we have a bottom-up approach where synergy exists and, thus, optimum productivity is more likely.

Achieving optimum productivity is critical for business performance. Wasted effort is lost cash flow and a negative factor for competitiveness. We can see that true productivity is gained through achieving communication excellence. In fact, communication excellence is the main productivity/cost reduction generator derived from CIM. A bottom-up CAPP system creates communication excellence at the work station and between functions. By uniting methods, time standards, and planning, we force communication excellence.

Bottom-up CAPP creates a logical, orderly set of communications of work to be done by and to all members of the business team. It creates a document describing what must be done. This document contains methods and instructions, sequence of operations, cycle times, and material-use sequences. This communication of who does what work, when the work is to be done, and what the work content is forms the basis for scheduling. If this work definition package is not based on accurate achievable cycle time, then the schedule is not a schedule but an estimate. Bottom-up CAPP creates data for schedules, not estimates. It is the accuracy of the schedule that sets winners apart from losers in today's competitive manufacturing world. We can see that a sufficient bottom-up CAPP system is a requirement for a successful MRP II system.

We will explore MRP II in depth in a later chapter. However, it is necessary at this time to state that if an MRP II system is supplied with accurate schedule time data, it will yield cycle time reductions, work in process inventory reductions, standardizations of processes, and dynamic scheduling capability. This is synonymous with productivity improvement and can only happen if the basic data, the methods/time standards/ planning, are accurate. Accuracy can only be achieved through a bottom-up building of information needed to complete the manufacturing tasks.

A bottom-up CAPP system contains the following eight elements.

1. *Methods and instructions.* These are the specific actions to be accomplished at the individual work stations. When combined with time standards, they become an operations plan.

2. *Engineered standards.* These are the incremental times for human and machine motions. Machine times are literally feeds and speeds converted to cycle time. For N/C machines, this would be the time for input of manuscript into the machine control unit. For human motions, times are based on the principles of motion economy. The specific incremental times usually are obtained from commercially available pretabulated times. MTM Associates is a typical example of a commercial tabulator.

3. *Operations planning.* This is the set of instructions to be carried out at a work station. It includes the methods and drawings, the associated times to accomplish the tasks, the sequence of events to be done at the work station, associated materials, associated tools, and usually an inspection requirement.

4. *Routing/sequence generation.* This is the compendium of operations planning in a work station by chronological order at the work station. Routing is the path to specificity selected work stations for the particular part. Sequence is the order in which the materials will be presented to a selected work station. This is the segment of CAPP in which primary and contingency routings and sequences are developed and stored in the database.

5. *Engineering change notice traceability.* Many product designs evolve even as the product is in production. It is important for future reliability considerations to know what version of the design was incorporated into each individual issue of the component. A good CAPP system provides the history of what models contain what version of the design.

6. *Group technology classification and coding.* The principle of sameness is employed to define like characteristics, both geometry and/or process based. Using a group technology code, it is possible to exploit this sameness to approach flow manufacturing efficiencies. A CAPP system uses group technology to group parts and processes at various levels to optimize planning work volume and sequencing.

7. *Variant or semi-generative planning capability.* Variant planning means using group technology classification coding to find "similar to" or exact match planning sequences. A new part design is coded, then a search is made for existing "similar to" or exact match planning. This significantly minimizes the planning cycle time and encourages standardizations within the functions. Semi-generative planning implies allowing the computer to select the process the part ought to be put through on the basis of the desired part characteristics. Allowing the computer to do the sequencing and routing implies some level of artificial intelligence (AI) or expert system, which is called semi-generative because AI is still an incomplete technology. Group technology classification and coding is usually employed as an initiator of semi-generative planning. One way of using code would be to link a part code with various process step codes and, in a semi-generative manner, select the order of operation for steps to be performed.

8. *Interfaces with other CIM modules.* This states that CAPP is an integrated member of the CIM strategy. Typically, we would see CAPP output going to the MRP II module, specifically the scheduling algorithm.

Figure 6.3 shows the schematic of a bottom-up CAPP system. The circled numbers refer to the eight items just described. Note that the CAPP system is divided into two basic activities: time standard activities and planning activities.

Time standard activities involve use of generic engineered time standards tables to create macrostandards that can be used over and over. An example of a macrostandard would be placing bar stock into a particular type of engine lathe head stock. Macrostandards are usually cataloged in a computer database and are called up as needed to fulfill specific job requirements. These macrostandards are the smallest elements of work used to create all the operations planning. Many CAPP systems also use group technology coding to identify these macrostandards.

It is interesting to note that state-of-the-art bottom-up CAPP systems use relational databases for managing data. This is done to enhance ad hoc query capability. Ad hoc query capability is the key for putting together many macros to form a plan. We must be able to mix and match macros, usually designated and focused by group technology coding, to do the job effectively. Older networking databases make the construction of the CAPP software tedious and cumbersome.

The first phase of the planning activities consists of consecutive buildup of the macrostandards to form linked sequences of operations. Standard operations development is the matching up of macrostandards to work at the work station. We can define

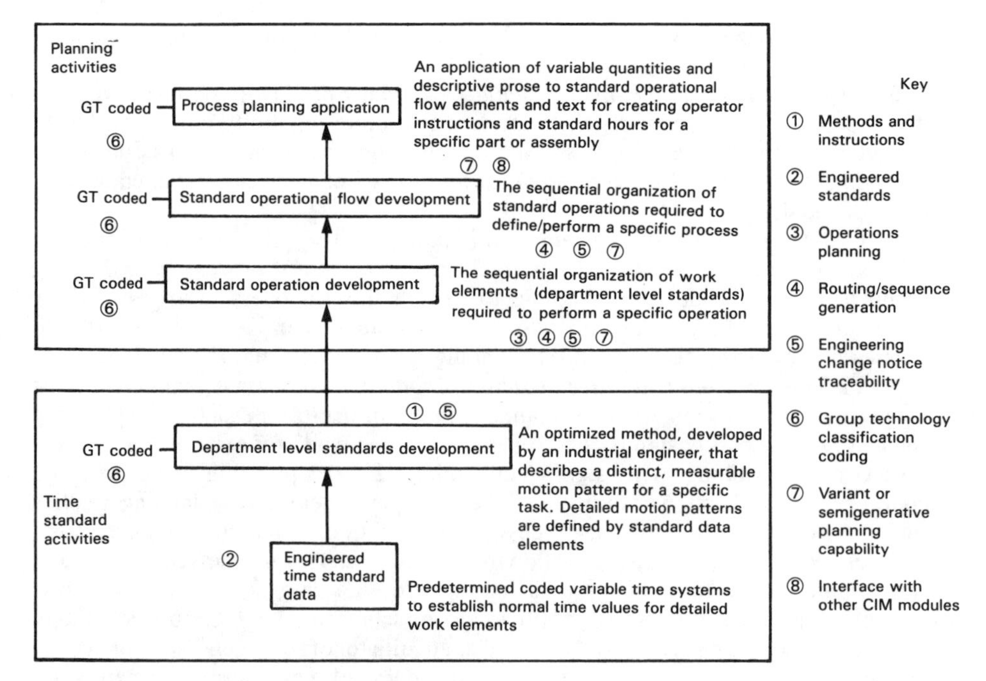

Figure 6.3 Bottom-up CAPP data build up.

this as an operation that can be done at a specific work station. So we now have a string of macrostandards linked into a standard operation, and that standard operation is conducted at one or several acceptable work stations.

The second phase of the planning activities for bottom-up CAPP is the linking of standard operations to form a routing and sequence over many different work stations. We are now creating flows. Here we can use group technology in a semi-generative planning mode to find the optimum flow based on linking the most time-efficient standard operations (made up of macrostandards) or we can use group technology in a variant mode to find standard operations that were used before to create a planning sequence for "similar to" or identical parts. The former always creates the most time efficient (by in-process/value-added cycle time) flow, while the latter only allows for standardization of the way similar parts are made. Standardization in itself is extremely valuable for overall factory efficiency; hence, variant planning is still a very viable alternative.

The third and last phase of the planning activity is the actual publishing of the CAPP-derived procedure (publishing here means either creating hard copy or storing in an electronic medium). Here we add quantities to be produced, thus going from a generic form to a specific form for an identified customer. We also add the specific text information dealing with the particular customer (e.g., paint color, trim selections).

Now that the bottom-up CAPP system has been described, we can see what a powerful procedure it is. We have created macroplans with traceable, definitive links to their incremental engineered time standards origins. This means the entire process tends to develop an optimized time to do a particular job. This is true because all the times are based on the principles of motion economy and, hence, are the most energy efficient for the task at hand. Contrast this with ordinary process planning, where a plan is developed and then an attempt is made to apply time standards (which are engineered time standards in a minority of the time owing to difficulty in application on a manual basis) to the output. This can be an approximation at best because of the nature of the top level output; the steps are already macros, thus defying ability to put accurate times on them. So, the times are not accurate. If the times are not accurate, then the second "know," know how much time it should take to make the product, cannot be fulfilled; hence, we are not operating optimally. This is the reason bottom-up CAPP is such an impressive productivity enhancer and is such an important component of any CIM implementation plan.

Closely associated with CAPP and vital for updating the common databases is data collection. Technically speaking, data collection is a component of the production and measurement control leg of the CAD/CAM triad. Since data collection is so closely allied with integrated CAPP systems, it is introduced here.

With modern integrated CAPP systems, we see that planning can be matched to and used to update the master production schedule. This means a methodology is required to keep planning needs synchronized with schedules. This is done through data collection.

Early data collection supported finance needs for periodic reporting. Therefore, it was not terribly sophisticated because sophistication was not necessary. A monthly update of status was all that was required. This was done by comparing the amount of direct labor hours turned into payroll versus the number of products made. From this, a calculation of unit cost was made, which in turn was compared with budget. The budget typically was based on experience plus perhaps a stretch factor and called productivity. Since most companies pay their factory labor on the basis of hours worked, it is very easy to compare hours worked with output, and from this comparison a whole host of efficiency and productivity measurements can be made.

Every month these measurements of actual performance compared to budget were presented. Finance reported the "objective" facts, and manufacturing made "subjective" analysis of those facts to explain variances from budget. Under this system there are always variances from budget because the budgets are not scientifically determined. The results are that the managers never know how good or bad the performance really is.

With the advent of CAPP, management had for the first time the ability to prepare budgets based on scientific analysis of capacity; thus measurements against that accurate baseline had real validity. The use of relativism, a seat-of-the-pants-type judgment, was no longer the best managers could hope for. Bottom-up CAPP led to true knowledge of factory capacity and, hence, to realistic budgeting. Since CAPP is capable of working in a dynamic mode to support scheduling, a reporting mechanism to get results fed back into the system on a real-time basis also is required. This led to the need for what we call data collection. Data collection in a CIM system is reporting to the scheduling module of MRP II, actual status of work in progress on a real-time basis. This is virtually always done by means of a computer terminal directly networked to the scheduling database.

With an accurate base for budgeting available, it becomes vital that the manufacturing managers have an equally accurate mechanism for decision making. To get at the causes for why schedule and actual performance are not in sync, fast and accurate data are necessary. Inaccurate data can lead to wrong conclusions, and slowly evolving data can leave the trail of the fault occurrence too far in the past for corrective action to be taken. Manufacturing needs to plan for effective actions and counteractions. This is only possible with good and timely data receipt. In most cases this means real-time data are required.

Data collection provides for real-time access to information. Basically, these systems

1. Record completion and starts of planned steps at the various work stations
2. Constantly compare work progress (status) with plan (schedule)
3. Do these tasks in an environment where the people making the data entries are virtually computer illiterate

A data collection system must be able to record the beginning and the completion of work as it occurs and must be able to compare constantly the actual status of the work with the original schedule. The system must also be able to keep track of updated versions of the plan. This requires real-time communications.

Terminals, whether stand-alone or integral with N/C equipment, must be able to issue clear instructions to the operators and let them query for clarification as necessary. Terminals must also lead the operators through the steps of entering data, to the point of verifying whether or not the input makes sense. Whether the terminals have local computing power or are limited to communicating with a remote host, they must provide ready access to the factory schedule database and must be able to update that database quickly and accurately. Because the terminals will be used by people who are not sophisticated computer programmers, the data collection programming must cover all possible contingencies and be easy to use.

A generic factory data collection system is shown in Figure 6.4. The diagram highlights the interaction between the operator at the work station and the computer system. The benefit of this interaction is that the operator becomes more knowledgeable about the master production schedule and the success and failures of the entire manufacturing component. Greater familiarity with the system allows the operator to take on more responsibility, which in turn has been shown to lead to improved productivity.

Figure 6.4 also shows that errors are expected and that provisions have been made for the operators to correct them. The error correction routine recognizes that operators are not necessarily computer literate and sets up a procedure that reviews input for obvious mistakes and protects the data from faulty input. These fault detection programs have to detect missing or incorrect information about times, quantities, part or drawing numbers, and other parameters specific to the factory and product. Error searches include instructions that step the operator through a correction routine. Data are finally

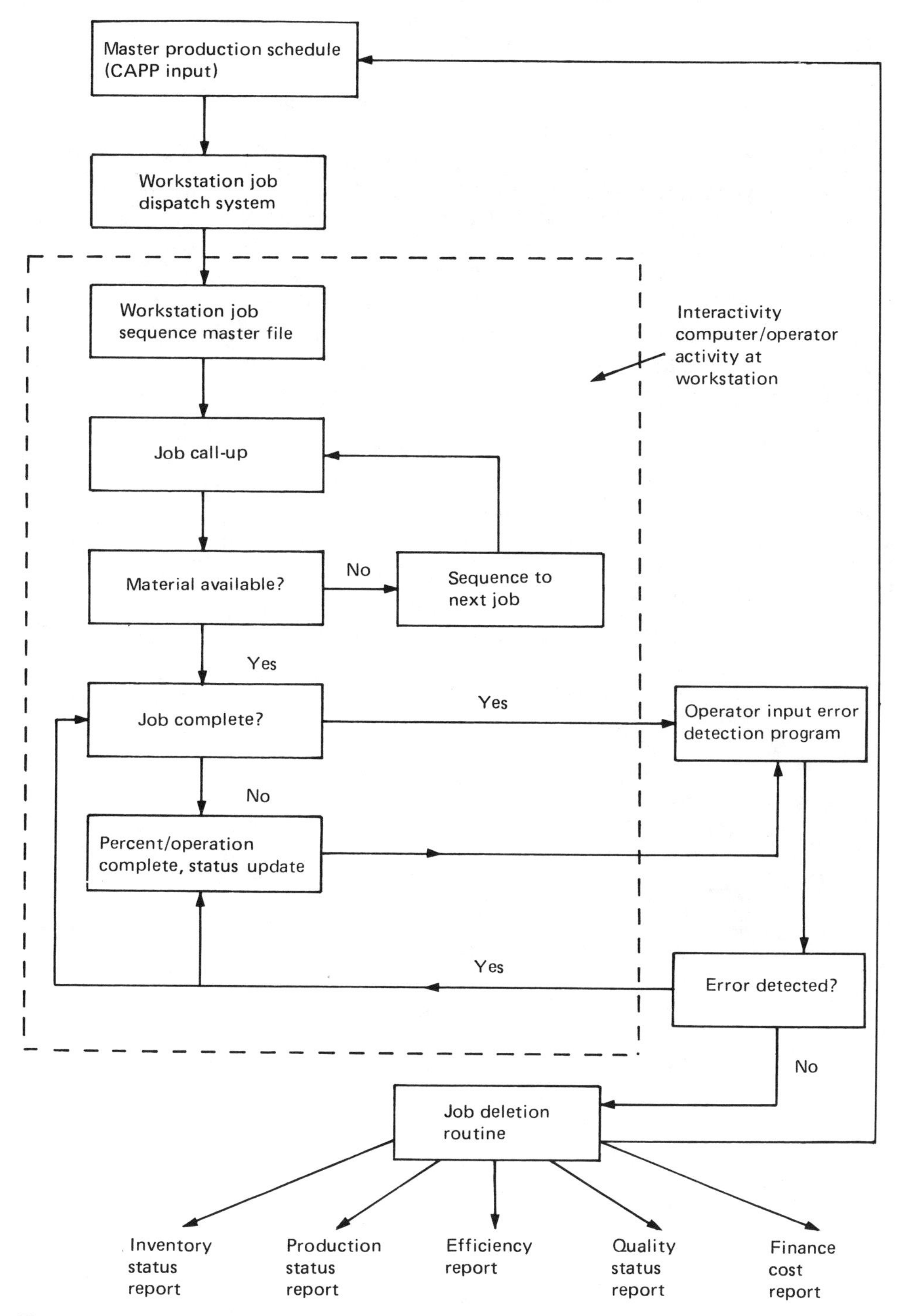

Figure 6.4 Generic factory data collection system. (From Daniel T. Koenig, *Manufacturing Engineering: Principles for Optimization,* Hemisphere, Washington, D.C., 1987.)

released to the master schedule and other databases by means of the "job deletion routine" once the error correction process is concluded.

It is important to recognize that a data collection system cannot tolerate the freelancing of operation sequences that is prevalent in manual scheduling and data gathering systems. If, for example, step three of the sequence cannot be completed because a tool is unavailable, it is not possible for the supervisor to order that step four be done instead. The system would balk at that unless the interchange is a bonafide contingency plan. This is good because all too often these freelancing decisions have resulted in poor quality products. Here we see an example of the system enforcing a much-needed discipline. Conversely, we see that data collection forces management to make sure that the sequences used in planning are in fact the actual sequences used on the factory floor.

This concludes the discussion of CAPP. We can now readily see that CAPP is vitally important to the CIM philosophy. It breeds integration and control and provides the basis for communication excellence.

Chapter 7

The Coming of Age of Statistical Quality Control Through CIM

Another example of the production and measurements control leg is the quality control system. Quality control is that broad range of activities devoted to satisfying customer requirements through the proper actions of the entire organization. Quality control can be thought of as the active conscience of the business. Its purpose is to keep customers contented with the knowledge that what they are purchasing will be in accordance with their desires and that the vendor is treating them with due concern for their special needs. Right away we can see that quality control is more than simple inspection and measuring of defects. It is a concept that demands that the customer be pleased. Therefore, it involves the entire work force, regardless of position. This is called "total quality control." The use of computers plays a major role in making total quality control a practical reality. We will look into the role that the computer plays in this process and particularly how it fits into the CIM philosophy. First, it is necessary to discuss the fundamentals involved.

In recent decades we have seen a growing awareness of the importance of producing products of impeccable and correct functionality, plus conveying concern for the customer's well-being. This is just good business sense that all too often has been ignored by inward-looking managers. In the past, quality control was very narrowly focused. It literally dealt with damage control (e.g., keeping defects down to an acceptable nuisance level). The key operational philosophy (particularly in American industry) was AQL, acceptable quality level. This philosophy legitimized the acceptance of less than perfection in anything industry did. What a defeatist attitude!

AQL procedures evolved around the mathematics of predicting the number of bad parts in any manufacturing lot. It codified, between buyer and seller, acceptable risk levels for accepting parts delivered based on statistical sampling plans. This system inevitably meant that products that did not function according to specifications reached customers. So, the quality control system's purpose was to keep the level of defects below the pain threshold of the customers. The system concerned itself with measurements of yield from the manufacturing process and equating that with what is acceptable. This means that tabulations of scrap and rework are done and compared to budgeted levels. Also, the quality organization functions as a complaint department processing customer claims and keeping track of warranty costs to fix defective products that eluded the internal detection systems. Since defects are expected, the burden of proof as to whether shop operations is doing its job adequately is left to the quality control staff. In other words, quality control is responsible for quality, and the producing functions are not. This unenlightened approach was shattered when Japanese competitors, devoted to American-developed but unheeded concepts of striving for zero defects, invaded the marketplace with excellently performing products much less prone to failure.

This led to the current concept of quality that uses the techniques of the old concept with a new goal of achieving zero defects. Yield as a sum of total production minus scrap and rework is still an important measurement; but, instead of comparing to an acceptable "pain" level budget, it is compared against the absolute level of 100% yield. Relations with customers change from an AQL concept to being one of not expecting customers to be satisfied with less than perfection. The fact that orders-of-magnitude improvement in product quality has been demonstrated means that customers have been trained to expect perfect products. We have increased the expectation level such that superior quality is now a marketing tool and not a lower level drain on the profit margin. Demonstrated capability to make products of high quality is now a key element in increasing the confidence of customers in the company and, hence, gaining a competitive advantage.

To achieve the higher levels of product quality, the quest for excellence has to be bred into the entire company. The Japanese have been very successful in this by emphasizing team approaches to solving quality problems, where everything the company does is fair game for analysis and change. This enlightened approach emphasizes people doing their jobs with their effects on quality always being the paramount consideration. Quality becomes a total process. We

- Design quality into the products from the start
- Strive to improve manufacturing capabilities
- Work with vendors to improve the quality level of materials purchased

These three areas definitely relate to the integrated concept of CIM. Design compatible with manufacturing capabilities and limitations of materials correctly implies use of common databases to store, retrieve, and analyze data. In the area of manufacturing capabilities, we are really saying that we are passing judgment on the process plan for making the product. To do this effectively, we have to analyze huge amounts of performance data and reach conclusions in real time. This has led to the development of computer-aided process control. Without such a system, real-time analysis is impossible, which significantly constrains abilities to improve manufacturing capabilities. We will discuss statistical quality control shortly. First, it is necessary to explain the process control system to see how data flow and how they are reacted to.

A process control system is a series of coordinated events in a manufacturing cycle that allows achievement of stated quality (zero defects) and production goals. Coordinated events are process steps linked to quality verification actions. These are conscious actions integrated into the method manufacturing chooses to make the product. This implies correctly that a process control system can be thought of as a method of who does what, when, with defined materials and equipments, and with what required checks (quality verification techniques). This is the traditional process plan with quality verification integrated into it. So it is not alien to the concept of making a product in traditional factories. It is simply placing a much-needed emphasis on verifying compliance with specifications and on the goal of producing 100% usable output. Let us look at a typical process control system diagram and see how it fits the CIM concept.

Figure 7.1 shows a closed loop system starting with quality plans dictating the required verification actions and then the action to be taken by the appropriate personnel in accordance with the quality plan. This action is then recorded by means of the data collection system, as described in chapter 6. This is the same data recorded to track compliance with schedule as referred to in the CAPP discussion and to be developed later in MRP II discussions. But, in this case, we are analyzing it in relationship to whether or not the component meets quality requirements. This means, does it comply with design specifications? Here we see the opportunity to use data collected for other purposes for this need. We once again see the concept of the common data base; thus, the CIM philosophy is applied once more.

Figure 7.1 uses terminology different than in the production cycles presented so far. We see QCR, IWR, and DAR. These refer to quality control report, incomplete work report, and defect analysis report, respectively. They are terms referring to the status of work and data that are easily recorded and compiled. A quality control report is a document, produced either electronically or manually, that reports on defective work with sufficient detail to allow analysis for cause.

An incomplete work report (IWR) is what the title implies, that is, work that should have been done but was not finished. Many times, operations are skipped because of changes in methods or plans. We discussed this in chapter 6, concerning the discipline needed to ensure that all required actions are accounted for. The process control system is the vehicle to uncover such glitches. Again, this demonstrates the compatibility of the quality system with other requirements centralized in a CIM common database.

The defect analysis report (DAR) is a generic compendium of any type of analysis that can be derived from the data. The most common DAR revolves about statistical quality control (SQC, often referred to as statistical process control, SPC). SQC was originally the most common use for computer systems in quality control, and we will discuss it later in this chapter. We can see that the input for SQC is obtained from a data collection system that is employed for multiple activities. Again (risking being redundant), this is in compliance with the integrating philosophy of shared data put forth by CIM.

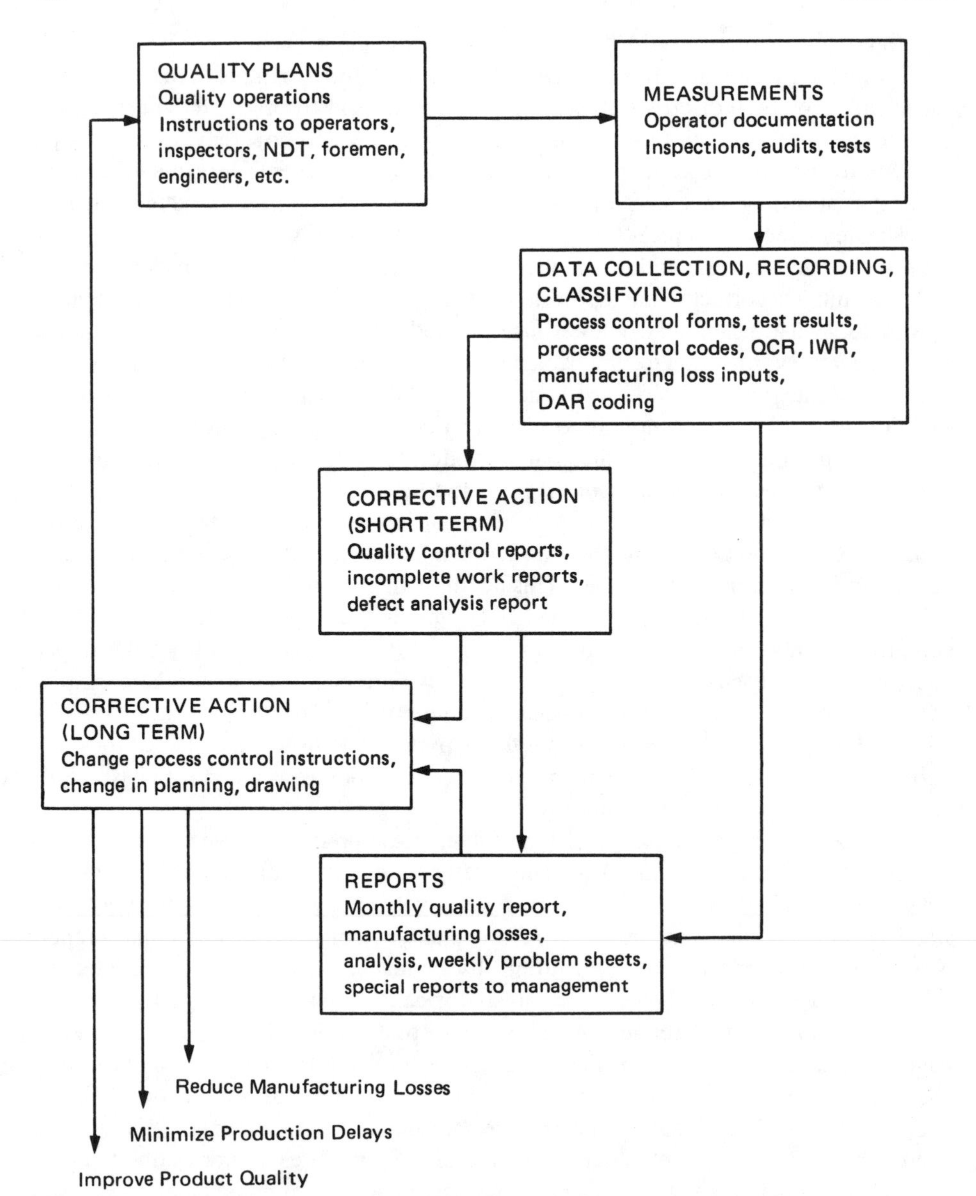

Figure 7.1 Process control system. (From Daniel T. Koenig, *Manufacturing Engineering: Principles for Optimization*, Hemisphere, Washington, D.C., 1987.)

Figure 7.2 shows the compatibility and relationship of the process control system with the closely allied scheduling and feedback system dictated by the manufacturing system. In fact, here we are looking at steps 2, 5, and 6 shown in one diagram. Integration offered by a common database is the key here. Because with computers we are capable of amassing and analyzing large quantities of data, it is possible to blur the lines of responsibilities, thus improving communication

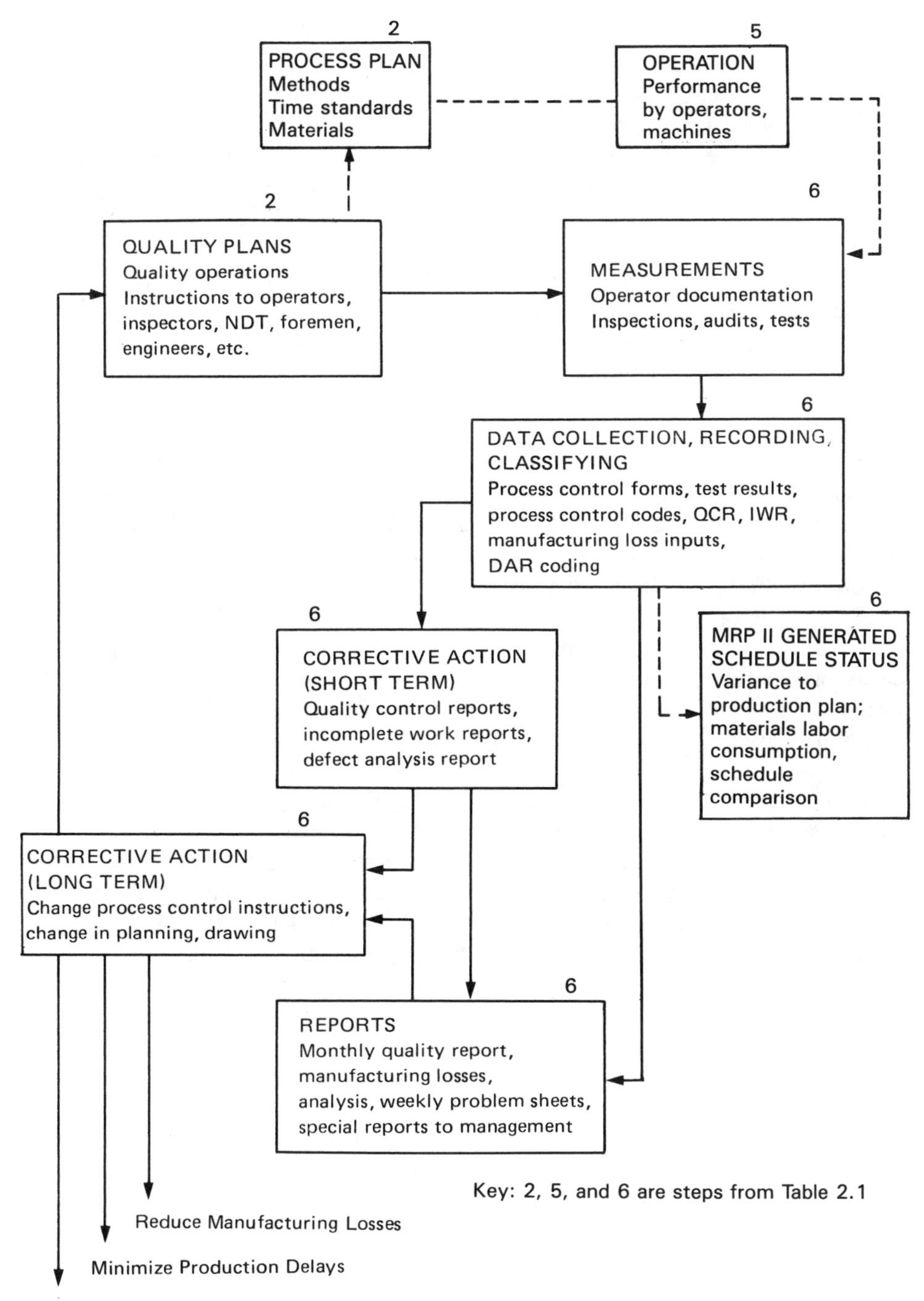

Figure 7.2 Process control system with related linkages to the MRP II schedule and reporting system. (Adapted from Daniel T. Koenig, *Manufacturing Engineering: Principles for Optimization,* Hemisphere, Washington, D.C., 1987.)

with results surely being improved overall performance. We are our brothers' keepers in more than the allegorical sense. By sharing real-time data, we react to it for our individual functional responsibilities, but we must do so in a manner that is compatible with our collective requirements.

The place where the computer makes the greatest impact is in the data collection, recording, and classifying parts of the process control system, as shown in Figure 7.1. We can tabulate and catalog large quantities of data. We can perform statistical inferences. We can chart trends to determine if a process is or is not in control. The quality plan states what sort of information is to be recorded and measured, and, as can be seen in Figure 7.2, it is integrated with the CAPP system (hence, the overall MRP II process). The data collection, recording, and classifying phases (also integrated with the MRP II process as in Figure 7.2) sort inputs and then categorize them. These data become the basis for reports and corrective action decisions. All this can be done without the benefit of a computer system. In fact, for generations it has been done by engineers and inspectors, who laboriously took data and applied probabilistic and statistic mathematics to infer trends and a state of the process (whether in control or not). This is tedious work and requires significant personnel effort and thus is difficult to apply to more than a few work stations at a time, the result being that applying statistical quality control to measure whether or not a process is operating properly is not done unless there is a significant problem (e.g., the process is thought to be out of control and data are needed to determine appropriate corrective action). Every competent manager agrees that it would be preferable to know if processes are drifting out of control prior to occurrence. But the cost of running these tedious inspections and doing the calculations for all work stations would be prohibitive, so it is not normally done. This, plus the fact that on a probabilistic basis the vast majority of processes are running properly, dictates that the cost cannot be justified. Conversely, if only one process out of a hundred were not operating properly, the defect costs could become catastrophic if not discovered soon enough. However, with a computer programmed to keep track of the data and to be continuously evaluating it, this dilemma can be resolved. It is now possible to do statistical quality control on a real-time basis and for all work stations. This is the true value of computer systems applied to quality control and the reason why we state that the computer has allowed statistical quality control to come of age. With this strategic and tactical background explored, let us now look at the statistical quality control techniques and see how the computer plays a large enabling role.

The basis of statistical quality control is the normal distribution of random events. On the basis of the theory, we take samples of an event and deduce whether the randomness is within parameters that are acceptable. We use the familiar parameters of the normal distribution to define the range of dimensions in which we are interested. The mean, or the average value of the distribution, is equated with the drawing dimension; and the standard deviations are equated to the plus or minus tolerances of the given dimension. The process is quite simple. We select a sample size large enough to be statistically significant, measure the pertinent parameters, and then calculate the mean and standard deviation by the following formulas:

$$U = \text{mean} = \frac{\Sigma x_i}{N}$$

$$S = \text{standard deviation} = \sqrt{\frac{\Sigma (x_i - U)^2}{N}}$$

where Σx_i = arithmetic total of measurements taken
N = number of measurements taken

Usually a $2S$ or a $3S$ spread is selected for the plus or minus tolerances. If we have a $2S$ spread, then, by normal curve theory, we are stating that 95.5% of all measurements will fall between these upper and lower values for the process to be properly established. If the $3S$ spread is selected, then the inclusive range expands to 99.7%. This is only true if the process used is operating in a random manner and is capable of operating in that bandwidth. If the process bandwidth does not match the $2S$ or $3S$ spread, then the process cannot meet the requirements, ever, and will have to be modified. Figure 7.3 demonstrates that, to use normal theory, we must first design a process that can yield results that are tighter than the specification limits.

A reverse application of this technique is used to determine if an existing process is capable of producing products within the specifications. Here, after ensuring that the process is operating correctly, we measure the sample to find its mean and standard deviations. The mean should be very close to the drawing dimension. If it is not, then the process is said to be skewed and the engineers have to adjust the process parameters so that the average corresponds with the desired dimension. Likewise, if the process

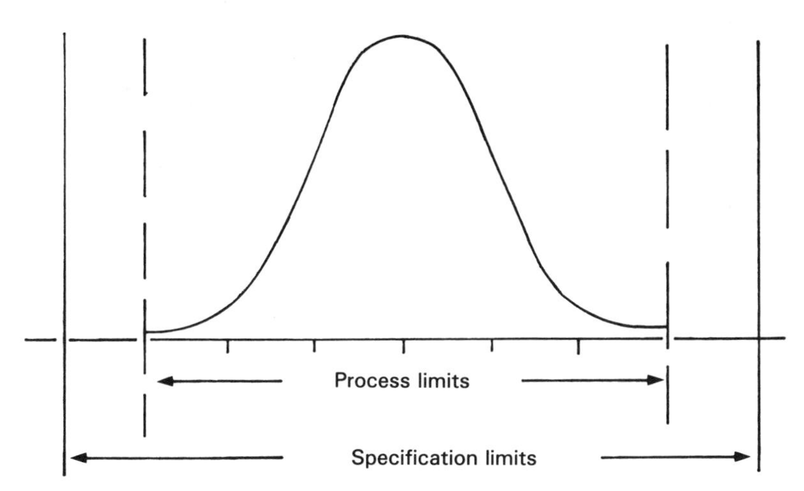

Figure 7.3 Specification limits versus process limits. Process limits should fit within specification limits. If not, process will produce rejects.

limits go beyond the specification, then the process parameters have to be tightened to produce a narrower bandwidth. This is a quite common application of the technique and is useful in fine-tuning the work station methods to produce within the required specifications.

Using a normal curve, similar to that of Figure 7.3, is a very awkward way of displaying data. It also has the drawback of only showing one set of data at a time. We are interested in data over time. Since normal curve theory is about random events, it is necessary to determine if events are truly random over a given period before any inferences can be made. Once we have the process and the specification properly matched, as in Figure 7.3, the process will continue to give good product if it is not deteriorating or changing. The determination of whether change is occurring is the heart of statistical quality control. We transfer results of sets of evaluations and process means to a chronological graph called a control chart. The control chart is a tool to detect nonrandom variations. Once a nonrandom variation is detected, usually while the process is still producing acceptable parts, the situation can be corrected before defective parts are produced. This is the power of the technique. We do not have to wait for defective components to appear before a fix is initiated, thus eliminating one potential for costly errors to be made.

Figure 7.4 represents a control chart. Each point represents a mean of a sample. The process average corresponds to the mean of the normal curve. The upper control limit (UCL) and the lower control limit (LCL) refer to the standard deviations, either 2 or 3. These also correspond to the drawing dimension and upper and lower tolerances if the specification limits and process limits are synchronized as described before. The example shown in Figure 7.4 is of the process not in control because it is not random.

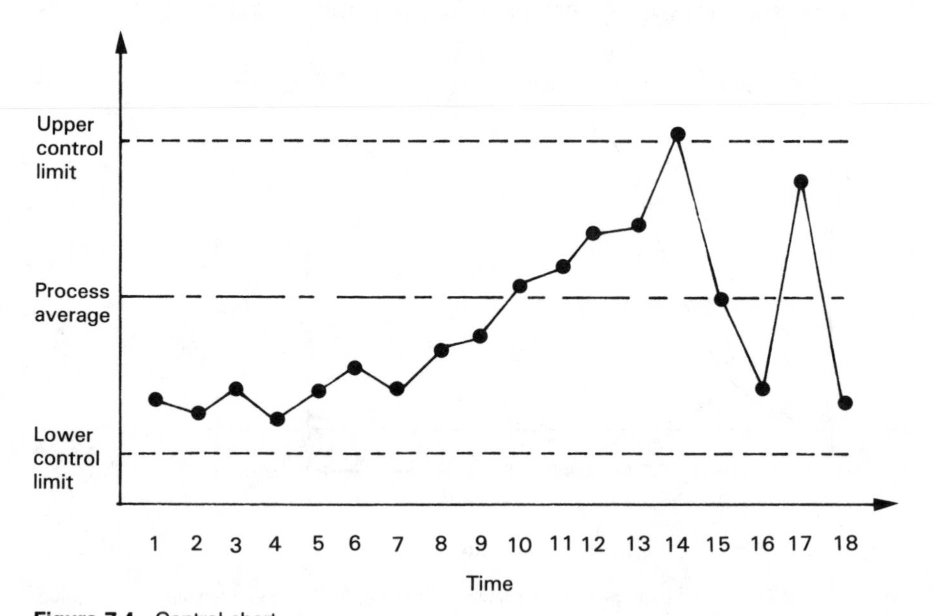

Figure 7.4 Control chart.

Note how times 4–13 portray a definite trend. This means the process parameters are gradually going out of control toward the high side, even though it is still producing acceptable product. This is like an automobile suffering from a wheel alignment problem. The car is still capable of being driven, but for how long? Note that at time 14 the process reached the border of being unacceptable. At this point a correction was made, and thereafter it appears that the measurements are random; thus, the process is back in control. We can see that using control charts is relatively simple. The problem is how to get the data, plot the data, and analyze the data. With a computer program keyed off process requirements we can collect data easily and write programs associated with the data collection system to plot and analyze. Let us look at this in more detail.

In a computer-assisted process control system, the computer is usually programmed to do the following seven steps:

1. Prompt the operators on what measurements to take.
2. Calculate the mean and the standard deviation.
3. Plot data on a control chart.
4. Determine if the sample is within acceptable limits and whether there is a trend toward unacceptable readings (lack of randomness).
5. Query the operator for causes if the sample is outside of acceptable limits.
6. Keep historical records of the process for reporting and documentation requirements.
7. Plot paretograph and other types of charts of defect causes and trends.

These are powerful analytical tools that are only available through the use of computers interacting at the work station and connected to a real-time data collection system feeding a common database. To do all seven steps just outlined manually would require additional support personnel. While a case could be made that the additional personnel would be well worth the cost for a specific or a few work stations, it would be very difficult to justify the costs for a firm's total factory. Some manufacturers do have a need for this intensive evaluation and normally can pass the cost on to the customer. An example would be a pharmaceutical manufacturer. Another reason why a manual solution is not advisable is the complexity of the process itself. Calculating the mean is a simple addition and division operation. The standard deviation calculation involves squaring, subtracting, summing, division, and taking square roots. While all these operations are well within the capabilities of an operator, they are tedious if done over and over during the course of an eight-hour work shift. Since there are so many calculations to do, it is virtually impossible to demand and expect that it be done 100% error free. If we cannot guarantee error-free mathematical operations, then the resultant data are tainted and are not to be trusted.

This is not the case with a properly written computer software program, so we can see that the only practical way to introduce SQC is by means of a computer systems mechanism. The benefits to be enjoyed if SQC can be implemented are well worth the cost of implementation, especially if there is a data collection system already installed for production control purposes. Even if SQC is the first CIM element to be installed, and has to shoulder the cost of the terminal installations, the computer, and the database software, it is still a very good deal indeed. The value of knowing that a process is drifting out of control before bad parts are produced makes a zero defect concept possible. This is a quantum jump in manufacturing's ability to satisfy customers.

SQC programs usually are programmed to look for nonrandomness. They use statistical inferences to make decisions as to whether or not randomness exists. They do it the same way an individual would, that is, by following a set of rules. When we discuss expert systems, a subset of artificial intelligence, I will use SQC as an example of that programming technology. Suffice it to say for now that the rules the programs follow are the same as those a mid-level expert in the process control area would use.

For SQC, the software programs typically track the seven most recent data points to test for randomness. They check to see if all the points are trending upward or downward. They evaluate whether the points are all above the mean or below the mean. The programs know that two-thirds of the points should be in the middle of the control chart when the process mean and the specification mean are in proper synchronization. If any of these conditions exist, then they consider the situation to be nonrandom and, hence, the process is out of control. We may ask, how does a program know that the seven points represent a trend even if the gross observations are as just stated? They do not, yet. The better SQC programs then do some type of regression analysis to see if a statistically meaningful trend slope exists. If correlation does exist, then a trend exists and randomness does not exist; hence, the process is not in control.

The early SQC programs ended right here. The software then put up a warning flag, indicating that the process was out of control, and showed the data points leading to the conclusion. The actions to take to find cause were up to the operator, supervisor, or engineer. All SQC did was to send out the alarm. The fix required expertise in the process to analyze for cause and implement a corrective action.

Current SQC programs are beginning to use expert systems to step an operator through a corrective action determination. Some of the steps would include a look at the process settings, a comparison to method requirements, and drawing of inferences from these observations. It is possible that the problem solution would be tested to evaluate the inferences. The program may also compare the current seven data points with historical data points of previously made parts from an identical component production run. If similar problems occurred, then the successful solution would be recommended. Here we see use being made of the expert system's ability to learn from experiences. In both cases, the procedure calls for developing a hypothesis of possible cause and then evaluating the hypothesis. Thus, the software follows the logic of the scientific method. It must be pointed out, to avoid confusion, that the computer is following a complex set of rules and, in accordance with the data it receives, is making action recommendations for the operator. The computer is not thinking, nor is it learning in a traditional human sense. It is simply expanding its software program if/then logic to an if/then probable then logic of an expert system. The use of an expert system SQC program makes all attempts at solving problems proceed in a manner that the competent expert would use. It cannot guarantee that the optimum procedure will always be employed, but the probability of approaching optimum would be greater. This is true because the randomness of having the best operator or engineer available to solve the problem precludes optimum procedure unless there is only one problem solver available

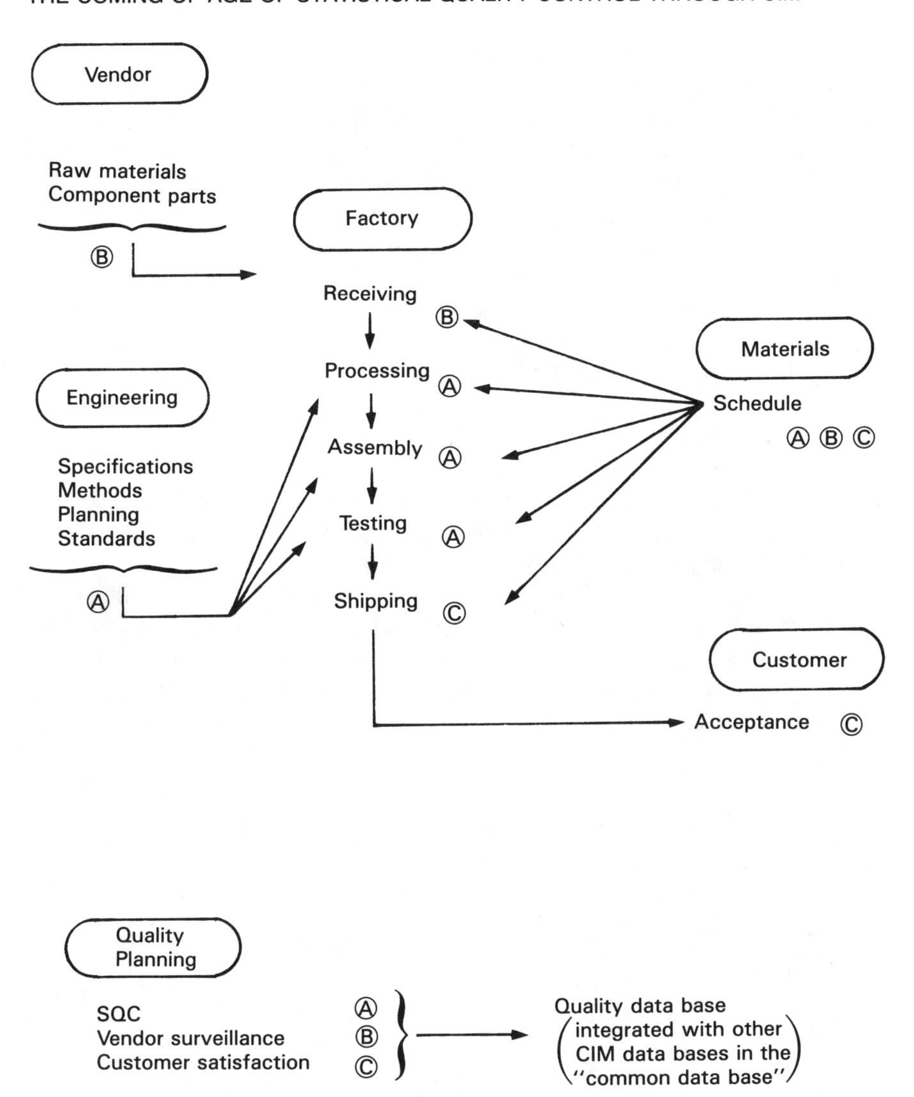

Figure 7.5 The total quality control database system.

and he or she always is the person doing the recovery work. This is the same argument put forth for the superiority of N/C machines over manual machines.

In keeping with the integration philosophy of CIM, SQC is often the dominant part of a total quality control database system (TQCDS). Most often, these types of databases use the relational-type of internal architecture database design because of the ad hoc query capability. From the previous discussion we can see that with SQC there will not be a majority of pre-set networks to be traveled by the data. Hence, there is a need for a more flexible database. The TQCDS interrelates with other databases to form a collage of common databases from which data can be exchanged, shared, and manipulated. Figure 7.2 is in essence the process control portion of the quality database. If we add vendor quality and customer satisfaction data (most often warranty experiences plus customer service records), we have a TQCDS database.

TQCDS, in the CIM environment, represents step 6 of the Manufacturing System. We see in Figure 7.5 how quality control relates to the manufacturing system.

It would be a very cluttered figure indeed if we showed all components of the common database on this figure. So, we emphasize the TQCDS superimposed on the manufacturing flow. The intent here is to demonstrate that there is a vital quality control portion of the CIM database and that it relates to and is compatible with the rest of the subsets of the common database. With this we can see that SQC is a part of the integrated whole. It does not stand alone. It receives data from the facilities of the data collection system and is part of the production and measurements control leg of the CAD/CAM triad. SQC comes of age because it is supported by powerful databases; thus, information flows, allowing the mathematics of statistics to be applied in a real-time sense. This real-time application makes it realistic to expect zero defects shipped to customers because processes are fixed before they produce defective parts. This is a real revolution in our thinking of how factories are managed. I believe it will lead to each and every customer rightfully demanding that any product purchased will perform as specified. This revolution in thinking, brought about through the capability of CIM, will cause quantum improvements in standards of living throughout the world. Amounts of funds similar to what we now use to buy error-prone products will in the future buy virtually perfect products, bringing more versatility and enjoyment to the users.

Chapter 8

Manufacturing Resources Planning and Just in Time: Two Compatible Philosophies Optimized Through CIM

A controversy has raged over the past 10–15 years over which is better: manufacturing resources planning (MRP II) or just in time (JIT). Proponents of each argue that theirs is the true solution to business's quest for optimization and that following their prescribed solution yields the ultimate success. I and many of my colleagues believe that both are significant and that both must be employed to make companies as profitable as they can possibly be. Furthermore, it is our contention that JIT and MRP II become powerful and logical techniques only if used together and under the umbrella of a CIM environment. In this chapter, we will explore the reasons for this and demonstrate the truth of these contentions.

We will start with working descriptions of MRP II and JIT, explaining what they are and how they work. After this it will be necessary to go into some depth on the theory of each. Finally, we will be ready to discuss the compatibility issue and how the CIM environment makes optimization of both possible.

MRP II started out as MRP (sometimes referred to as little MRP in reference to big MRP, MRP II). This had to do with computer-enhanced materials ordering and inventory control systems, very advantageous in enhancing the speed and accuracy of issuing raw materials to factory work stations. It immediately became apparent that linking materials with production demand schedules could optimize the flow of the product as it is being constructed in the factory. This could be done in such a manner that material queue times could be minimized (e.g., have the material show up only when needed), and ultimately the amount of material needed throughout the factory at any one time could be reduced. This is an optimization technique that allocates identified sets of materials (sometimes called kits) to specific jobs as they go through the manufacturing process.

Since it is possible for the computer to keep track of large numbers of kits, it in a sense reserves or mortgages materials for specific jobs in time-order sequences. Linking these sequences with a production plan based on customer need dates allows management to release and track orders through the shop very accurately. Prior to releasing orders by means of the kitting process based on the production schedule, it was necessary to obtain supplies on a gross basis dependent on the number of orders expected to be shipped to customers over the selected time period and by having the gross amount of inventory on hand at the start of the period to support production. Obviously, the kit will result in less extra materials on hand at any point in the production period. This results in large raw material reductions and reductions in material needs for work in process and, hence, lower operating costs.

In the early 1970s, theoreticians picked up on this fact and realized that information necessary for success in MRP was similar in nature to that necessary to computerize production schedules and that it was desirable and possible to link work station schedules with materials use schedules. In fact, all manner of schedules affected production through the common database concept. Hence, MRP II came into being, as a concept first, then as a set of independent programs, and finally as we know it today, as a component of CIM. So, MRP II started out as a materials optimization technique and grew into a total business scheduling optimization technique. Table 8.1 summarizes the benefits of this type of technique.

Table 8.1 correctly implies that massive amounts of data can be manipulated to make comparisons to a predefined plan, and actions can be taken as a result of current information very quickly, if necessary. No mention is made as to whether the schedule is a labor schedule or a parts schedule because it is irrelevant. Both can be merged into one, if desired, or they can be kept separate. We see that MRP II is a grand scheduling mechanism. It is usually a multiple database system maintained as part of the CIM common database. We will discuss the component databases necessary for MRP II later.

JIT did not start out as a computer-based tool to enhance productivity. It is a productivity enhancer based on a simple proposition that all waste in the manufacturing process must be eliminated. It has since become a computer-based concept simply

Table 8.1 Advantages of CAD/CAM Scheduling

	Access to common database
1.	Handle large permutations easily
2.	Restructuring of workstation loading and high level master production scheduling accomplished quickly
3.	"What if" scenarios can be examined
4.	Real time information easily available

Note: Adapted from Daniel T. Koenig, *Manufacturing Engineering: Principles for Optimization,* Hemisphere, Washington, D.C., 1987.

because the computer is so very useful in support of the actions required by JIT to carry out the basic philosophy.

JIT theory states that waste can only begin to be eliminated if the push production control system is replaced with a pull production control system. Since the emphasis from the start is on production control, we very early on had unfounded comparisons with MRP II. JIT states that waste must be eliminated, and a very large component of waste is bloated inventory levels. Therefore, we must find a way to minimize inventory levels. If you do this without the analytical capability of the computer, then it is logical to assume you would conceive a system that will not let material move or be used until it is necessary. This, as we all know now, is what Toyota did. Toyota is generally recognized as the first major proponent of JIT. They instituted a backward scheduling technique that started with the desired ship date. They had to know when the product needed to be at final assembly and before that when it needed to be at the subassembly levels and so forth, back through component part manufacturing. Ultimately, it means determining precisely when the raw materials should show up at the receiving dock. This, in itself, is not unusual or unique. Toyota did not have sophisticated computer scheduling algorithms with feedback capability but did not want to be burdened with excess inventory required by its relatively gross scheduling to ensure the factory did not run dry. So, Toyota arbitrarily said hold everything at the preceding work area until the succeeding work area required it. This is the pull system, as compared to the push system, which makes parts to schedule and ships them to the succeeding work station whether or not they are really needed.

To put the "rule of law" into their pull system, Toyota even went one very effective and famous step farther. They denied permission for the preceding work station to even make their components until a specific demand was placed on them. This is the *kanban* system: A specific request for a part to be made has to be received by a work station before it can start. This, in turn, forces the work station to stop its suppliers from sending work to it before it has a need.

As can be seen, this does reduce work in process inventory. It also puts the burden on the producing work station to perform correctly and efficiently because it has no buffer stocks to fall back on. So, if efficiency is by necessity increased, and correctness of product produced is also by necessity increased, then waste is minimized and profit is maximized. This, in essence, is what JIT in its barest and purest form is. We will never know if the eliminate waste concept/theory drove the development of the philosophy or if, as some suppose, the theory was derived after the success of the pull system was demonstrated. However, it does not matter; what does matter is that application of the concept reduces cycle time, inventory, and waste.

The trouble most Westerners have with the Toyota experiment is that it operates on a brute force level of manual *kanbans* and requires very close cooperation between vendors and customers. Let us look at the cooperation first.

Upon investigation it became apparent that the Toyota relationship with the vendors is not at arms length as much as it is in the United States and in Europe. In Japan, it is common for a major manufacturing firm to have an ownership relationship with the vendors, thus affecting how the vendor responds to delivery requests. This is not to say that *kanban* will not work with truly independent vendors, but it imposes more obstacles, the foremost being why a vendor should hold material at its own cost for its customer. This certainly does not optimize the vendor's cost/profit ratio. In the United States, some of this obstacle has been removed by implementing good management practices in the vendors' plants to create win-win situations. Another factor concerning vendors is that in Japan suppliers for specific industries tend to be grouped geographically very close to their major customers. Also, suppliers tend to supply only one major customer (probably owing to the minority ownership position major firms hold with their suppliers). This is not true in the United States, thus creating another obstacle toward implementing JIT. For these reasons we can effectively minimize the *kanban* philosophy utilization once we reach the receiving dock.

The use of *kanban* itself must be deemed primitive, and most companies have reached that conclusion. For a very simple process it is practical to yell "Hey, Charlie, make me five drimpoles by 11:00 A.M.," and Charlie will then do so and thereafter sit down on his stool and wait for the next "hey Charlie" yell. In practice, we must anticipate the pull needs and have parts ready when they are requested. This is like a nurse having the scalpel available for a surgeon the instant before it is requested, that is, it happens through practiced teamwork. In the CIM environment, this practiced teamwork comes about through application of databases to the goals of *kanban*. This is what has happened in practice as JIT was implemented in the United States and Europe. If we look once more at Table 8.1, we see that the benefits of real-time information can be used to adjust schedules to affect "hey Charlie" calls in practice. We use the principles of the pull system to load work stations. We do this in conjunction with MRP II scheduling overall and use *kanban* principles to affect daily and hourly schedules, as needed. This is the first of many examples that will be made of the compatibility of MRP II with JIT.

The JIT theory of eliminating waste means producing exactly what is needed when it is needed and getting the component part to where it is needed when it is needed. Robert W. Hall, in his book *Zero Inventory*, points out six prominent parts of the definition of what eliminate waste means (see Table 8.2).

He and other proponents of JIT state that these six parts or axioms, if followed, will result in the elimination of waste and allow the benefits to be achieved. If we look at these axioms, we see nothing startling. But, what we do see is a surprisingly simple restatement of the concepts of industrial engineering principles for operating an efficient enterprise. I know of no managers who could argue with any of these premises. In fact, these premises are so intuitively obvious one has to wonder if there is anything revolutionary at all about JIT. I do not think so. JIT is called by its promoters a

Table 8.2 Just in Time Theory: Goals

1. Produce the product the customer wants
2. Produce products only at the rate the customer wants them
3. Produce with perfect quality
4. Produce instantly; zero unnecessary lead times
5. Produce with no waste of labor, material, or equipment
6. Produce by methods that allow for development of people

Note: From Robert W. Hall, *Zero Inventory,* Dow Jones-Irwin, Homewood, Ill., © 1983, used by permission.

fundamental way of thinking to transform overall manufacturing in the simplest way possible and to generate new and original techniques for doing so. I cannot go that far. I believe JIT is a definite reawakening of industrial engineering principles in that it abruptly and sometimes brutally forces us to do what we should be doing as best we can and not be encumbered or sidetracked by the elegance of the tools we use. We should always be striving to optimize processes within the factory and external to the factory to create the best products and highest profits possible. Not to do so is almost a violation of the professional code of ethics. What JIT does is focus engineers on doing what they do best, that is, to strive to optimize the manufacturing system. In all honesty, this is something good engineers have been doing since the onset of the industrial revolution. So, to traditionalists, JIT is simply saying do the job properly, and it is nothing new at all. JIT, then, in my opinion, is not a theory or a strategy but a tactic in carrying out good engineering practices.

Let us now discuss these six axioms and see how they relate to a CIM environment. Keep in mind that JIT was devised to optimize production in a manually controlled factory equipped only partially with N/C machine tools. If JIT as a tactic is to have a future, these six axioms have to be applicable and practical in a CIM environment.

1. Produce the Product the Customer Wants

This is so intuitively obvious, it is almost insulting that it is necessary to list as one of the axioms of JIT. However, we have found that managers and operators have become so entranced with the nuances of running an operation well that too often processes are run because of the "high" it gives to those in charge to be doing something they consider to be elegant, regardless of whether there is an immediate requirement for the output. Here we see people getting caught up in specific work station efficiency and rationalizing that the output will be needed (e.g., the yearly output). In these cases, those responsible fail to realize that they are making waste if a product is not immediately needed and that the effort expended could have been used more productively elsewhere. The guideline to keep in mind here is that work station efficiency/effectiveness only has relevance when it is compatible with overall system efficiency/effectiveness. Producing what the customer wants means supplying only the demands of the next work station.

The fact that we produce only what is requested is compatible with CIM philosophy. CIM requires communication excellence. I can think of no more effective means of communicating desires and requirements to producers than through the common databases demanded by the CIM solution.

2. Produce Products Only at the Rate the Customer Wants Them

This axiom logically follows the first. It means do not create a hoarding mentality but assure the customer that the product will be available at the time required and in the exact quantity necessary.

This, again, can only be optimally ensured through a CIM solution. It is necessary to keep track of succeeding work station (customer) needs so as not to under- or over-produce. The best way to do this is through a dynamic feedback control system production schedule, as offered by MRP II, and it can only be done through the CIM solution. We can monitor progress of all work stations carrying out the dictates of the strategic plan and thus speed up or slow down the preceding operation to achieve optimum use of materials and labor. Contrast this with the manual *kanban* system first thought to be the heart of the JIT tactic.

Kanban depends on an elaborate messenger system releasing preceding work stations to do work, the "hey Charlie" method. Lacking a computer monitoring and control system, it is better than nothing and most likely superior to other manual scheduling systems. But, with a manual system, we find that once an order to produce is given, it is rarely ever countermanded. With manual modes, we find that there is very little capability to reverse decisions once they have been set in place. With computer-driven systems, it is possible to program test hypotheses and "what-if" scenarios that are set to trigger reversal actions instantaneously if the situation demands it. This gives us the ability to reverse the "hey Charlie" order if it is appropriate, thus making it possible to achieve higher orders of optimization than is possible with manual *kanban*-like techniques. This is analogous to the computer-driven decision analysis required for space vehicle orbital reentry. Without such systems, space flight might not be possible because human calculation time and decision time is usually not fast enough to affect the desired outcome. However, if given the time to preplan and instruct the computer what to do in each of several probable situations, the computer will react correctly virtually all the time. The same is true in a dynamic production environment where all the variables from machine breakdown to nonstandard materials being encountered can affect the scheduled performance. Once the schedule is affected, it has a cascading effect on the preceding and succeeding work stations. In the optimum scenario, this requires instantaneous reformatting of schedules for all work stations. This is something JIT espouses but cannot achieve through the manual *kanban* system. It can be achieved readily through a dynamic scheduler in an MRP II system working in accordance with the JIT axioms.

A good example of this axiom being applied lies in the area of preventive maintenance. We say we want to produce at a desired rate. One thing that can keep a factory from producing at a desired rate is unscheduled stoppages of processes, that is, the breakdown of equipment, usually due to poor maintenance activity. A CIM philosophy, like JIT, implies properly operating equipment.

JIT axiom theorizes that not having equipment available when required breeds waste. Indeed it does! The equipment must run to make the product. Otherwise, waste is generated by means of paying operator salaries and not getting any value added in return. JIT emphasizes preventive maintenance because equipment is not expected to run unless a requirement to run is generated. Therefore, it must run when demanded and must not run when no demand is evident. Otherwise, waste is generated.

To be able to run on demand, time must be set aside to make sure that all equipment is capable of responding to the call, like the city fire department. We would be upset greatly if the pumper could not respond and our house were burning. The question then comes up: When can we make time for preventive maintenance? The JIT axiom theorizes that since equipment only runs when a demand is placed on it, and not continuously as in a push system, there ought to be ample time to do preventive maintenance. The key here is the emphasis on preventive maintenance. We do not wait for breakdowns but do logical things to prevent breakdowns from occurring. Since we are no longer concerned with individual work station efficiency in a JIT scenario, then there is ample time to do the preventive maintenance. This has proven to be true in real life examples.

3. Produce With Perfect Quality

This means doing it right the first time. It implies that the design is a producible design for the specific factory and that the methods for producing the product are well thought out. If we can produce with perfect quality, then there is no need for reserve inventory. In fact, early proponents of JIT said to reduce the reserve inventory levels to make managers and operators rise to the occasion and to find better ways of doing things. This is the school that says the way to learn how to swim is to be thrown into water ten feet deep and at least twenty feet from safety, a risky process that relies on necessity being the mother of invention. I believe proper application of the industrial engineering process (e.g., the use of the scientific method) will yield more formidable results, swifter and with less risk.

Producing with perfect quality is a laudable goal and is supported by all responsible managers. The goal may be unobtainable over the long haul, but to have a goal that is any less inspiring is in a sense accepting defects and the rationale that goes along with them. As was discussed in chapter 7, we see that zero defects has to be the objective the organization strives for. If it consciously does so, then the defect level will be far less than that for those who resign themselves to the fact that defects to some degree are inevitable.

The premise of JIT of perfect manufacturing is obtainable; therefore, we can achieve optimum efficiency. I am not so naive as to believe that anything human beings do can always be perfect, but I do know that objectives have to be set to achieve perfection and, if they are, then for longer periods of time defect-free manufacturing can occur. Original thoughts of JIT implied that near perfection occurs through awareness of the need and intensified effort by all involved. Anyone who has ever led a team effort, be it Little League or charity drives, knows that intensified performance can only be maintained over a short period. People quickly tire of the novelty, or just plain get fatigued, or the crisis imposed to get extra effort becomes a noncrisis, or for whatever reason performance wanes and we are back to pre-hoopla performance. JIT tactics cannot be long-term strategies unless there is a way to institutionalize improvements.

The only way I know of to institutionalize change in factories is through the use of the CIM philosophy of communication excellence by means of common databases. To produce with perfect quality, we need excellently conceived CAPP systems supporting a dynamic scheduling capability, as suggested by MRP II, and a dynamic evaluation of results in progress capable of being provided by properly designed SQC. As we can see, the JIT idea can come about in a permanent way by instituting components of CIM. The computer does not lose interest, is not distracted by the new problem of the month, and can continuously evaluate data to give optimum results. This is not possible with JIT envisioned as a manually implemented tactic.

4. Produce Instantly: Zero Unnecessary Lead Times

This means having equipment ready to respond as the requirement becomes known. It does not mean having "safety stock," material on hand not previously assigned to a job order. This would defeat the intent of the *kanban* philosophy. So, we can say that this axiom applies in a pragmatic way to in-process reactions to demands (i.e., the component is already somewhere in the schedule but its start date has not been previously assigned). Perhaps it is on the general manager's master plan for the period in concert with his or her sales goals. Once the start date is assigned, we want as fast a start toward satisfying the demand as possible. By implication, it probably also means that if materials have to be obtained that they be obtained in an expeditious and as least costly manner as possible.

In the manual JIT approach, equipment is waiting to be used because we do not put any premium on running machine tools continuously. Long production runs are not looked on favorably unless there is a definite immediate need for the output. So, the pull system operating under the *kanban* method is more likely to be able to respond to the produce instantly command than a push system factory. Quick set-up times become a desirable attribute, and flexibility is sought as if it were the pot of gold at the end of the rainbow.

CIM, by means of flexible automation, is part of the answer to the desire to produce instantaneously. The flexible machining center, with its ability to do single-part orders as well as multiple part runs simply by installing a different software program in its MCU, is currently the only answer to the command to make parts with short lead times. This type of N/C machine requires very short set-up times, and thus it is extremely flexible when compared with state-of-the-art hard automation and is almost as efficient at the work station evaluation level. Many manufacturers have tried this approach and have achieved partial success. The reason for only partial success is that they did not go far enough. They only addressed the machine/process control leg of the CAD/CAM triad and did not implement the necessary components of the other two legs. They did not implement the CIM philosophy. When they do implement the other two components of CIM, they find that the JIT process works just fine.

If the flexible machining centers are coupled with good scheduling systems that change dynamically with needs, and can find the proper methods, and in turn the materials necessary can be dispatched quickly, then we can quite readily obey the dictate to produce with zero unnecessary lead time. What I have just described is hooking up the machine control unit (MCU) with a CAPP system, an MRP II scheduling program database, and a material handling system such as an automated guided vehicle (AGV) dispatched from an automated storage and retrieval system (ASRS). These are all components of the three legs of the CAD/CAM triad and derive their required information from the CIM common databases. Once more we see that this axiom of JIT really requires a CIM solution to be practical for the long term.

5. Produce with No Waste of Labor, Material, or Equipment

This means manufacturing at optimum conditions. To fully comply with this axiom, a factory has to do everything correctly. Jobs have to be planned perfectly, which means engineered time standards have to be set for a method conceived through expert use of the scientific method. Material has to be on hand only when required, and it has to have a 100% pedigree (e.g., it exhibits perfect compliance with the design specification). Equipment has to be in textbook-perfect condition. This means machines are perfectly maintained and tooling is superb, always producing parts to specification. No waste means that the work force is trained to perform as a well-drilled team, a team that knows precisely what to do and can react properly to unscheduled aberrations to routine. Is this possible with the original concept of a manually driven JIT? Perhaps not, but, as with any ideal to strive for, to approach it will give better performance than if the organization admits defeat beforehand. The problem is how do we institutionalize improvement so that the factory does not drift back to pre-JIT days? Again, the answer lies in the use of CIM principles.

One definition of producing with no waste means driving toward a factory that can produce single items with the efficiency of mass production. This means absolutely accurate scheduling systems are necessary, and they have to be in real time. Batch scheduling, the hallmark of manual systems, will not work because it is too slow and cannot account for all permutations underway. So, we need to introduce relational, on line systems to cope with the complexity. This can only be approached by means of MRP II systems supported by bottom-up CAPP. The manual *kanban* system, a brute force technique, bogs down because there is no way a manual system can achieve the 99+% accuracy that is required. To be error free, the only way conceivable is through a CIM-conceived common database philosophy.

Another way of defining produce with no waste is to optimize throughput time. It is a sad fact that in job shops 95% of the portal to portal time of producing a product is queue time and move time. Only 5% is value-added time. For flow manufacturing, we achieve on average 30% value-added time. These are not satisfactory grades, and the JIT axiom encouraging us to produce with no waste is well founded. It is deserving of finding a way to do it. I am afraid that the proponents of JIT stating that we should pay greater attention to detail in operating the manual system will not result in substantially better performance. We get back to creating institutional change to have any true and lasting impact. In the domain of material movement, to reduce this non–value-added time we must rely on computer-controlled modern material handling systems, systems that link automated guided vehicles to the machining cycles to reposition work pieces for the processing tools. This eliminates stops and the manual recalibrating and rethinking of steps people usually do. If we further link dispatch of material to the scheduling algorithm, we further reduce manual dwell time. So, with material movement ratcheted down by coordination, planned by people but carried out with precision by computers, we can institutionalize change. We can create a situation where we are reducing waste. I guarantee, not all waste but substantial amounts.

Still another elimination of waste is to design the process to use as much of the raw materials as possible in the creation of the product. We can conceive of how this should be done by requiring that all steps be planned to use the optimum approach. The method to be used must require the least amount of effort (another way of stating conservation of energy). We all know this incorporates the use of engineered time standards. We also know how to do this in a theoretical sense. Taylor, the Gilbreths,

Gantt, giants of industrial engineering principles, theorized how to accomplish this. Unfortunately, the work is time consuming and thus not practical for any but the most critical of manufacturing processes. JIT espouses the use of these techniques or at least their rough approximation cousins. Fine and good, but how do we institutionalize this change without breaking the financial back of the implementing firms? Certainly not manually. Some of us have had the onerous duty of trying to install engineered time standards manually across an entire factory. I found it took more than 5 years just to cover the five hundred or so work stations in a thousand-person factory, and even then we had only done a cursory job. So, I marvel at proponents of JIT who simply state "go do it." The only way to approach it is, once more, through the CIM-based philosophy of using common databases. In this case, we are using the bottom-up CAPP approach, which will be used to support an MRP II schedule development and a data collection system for feedback analysis. Tying this up front with the bill of materials provided by design engineering, we ensure that the factory is making the product as called for by the specification. Again, we can see how change called for by JIT has been accomplished and institutionalized by the CIM philosophy.

"Eliminate waste" is a noble sentiment, one I firmly support. However, I cannot agree with the glib pronouncements of many irrational supporters of JIT. I see no way of implementing this axiom of JIT without its being fully integrated within the CIM philosophy. JIT concepts certainly point out the way to go. But, they are pragmatically impractical unless they are tied with and become part of CIM implementations.

6. Produce by Methods that Allow for Development of People

This means use all the brain power available to solve all problems, of production in particular and the business in general. In all too many factories, and even in service businesses, we categorize people into either thinkers or doers. This is not good. All people, regardless of current job responsibilities, can make significant contributions in both areas if encouraged to do so. This aspect of JIT does not need a computer system to institutionalize. It simply takes an enlightened management to see that having all employees focused on solving the firm's problems is much more apt to result in finding useful solutions than if problem solving is relegated only to a management class. The arithmetic is simple. If a firm has 100 people employed, and 10 in the management ranks, then it is more likely to gain a competitive edge if all 100 focus effort on solving the problem than if only 10 did.

In practice, this development of people manifests itself in activities such as establishment of cross-training programs for operators, off-duty skills enhancement programs, and quality circles for problem solving. There are many more, and it would be worthwhile to discuss them, but they do not relate directly to CIM applications. Therefore, I will defer to other authors on this subject.

This discussion of JIT as it relates to CIM shows how JIT is applicable in the CIM solution. Therefore, it is a tactic with a future. I believe JIT plays a significant role in directing how the integrating concepts put forth by CIM should be implemented. It is easy to see that we should be striving to implement pull systems. JIT shows us this, and we should then devise our MRP II scheduling systems to effect that desired end.

MRP II can just as well optimize scheduling of push systems. However, we can see through the discussions of JIT theory why that would not be the proper way to go. Indeed, MRP II as part of the overall CIM concept must be made to be totally compatible with the axioms of JIT. If MRP II is not, then it is not a true CIM solution.

We have covered the theory of JIT and find it to be a process-enhancement tool, a tactic. We will now switch our attention to the theory of MRP II, and we will see how it is a procedure-enhancement tool, a strategy. I think we already see that the two are different but compatible. Our purpose will be to demonstrate this conclusively.

Let us start with little MRP, material requirements planning. This is a computerized ordering algorithm used for purchasing and consequently for inventory control. MRP is based on the bill of materials (BOM) developed by design engineering. The BOM is a list of materials in a "goes into" format very similar to that found in instructions for assembling model airplanes, cars, or ships. It lists all parts and subassemblies in the chronological order that they get put together. Sometimes manufacturing has to create a manufacturing BOM to factor in process steps that may not be apparent through a simple "goes into" pattern. An example of this would be an intermediate paint step necessary to coat the surfaces while they are still accessible. This, then, becomes a phantom level in the "goes into" chronological pattern, not necessarily a part or assembly engineering can describe but a process required for successful assembly.

The manufacturing BOM becomes the basic tool for organizing when materials will be needed. If we know the cycle time for making each part and doing each assembly, then we have the basis of a schedule and can produce a set of materials requirements need dates. This is fed into the MRP database, which compares need dates with purchasing lead times and sequences releases of orders to prequalified vendors. This way we can order materials to arrive in a JIT mode, thus minimizing inventories throughout. This is the basis of MRP. There are, as you would expect, nuances for successful or practical practice, but, in essence, this is all there is to it.

One of the more irksome questions that comes to mind with such a system is what should be purchased and what should be made in the factory. This mundane power play between the production manager and the purchasing manager as to who would have the biggest "empire," I believe, led to MRP II. With MRP, the purchasing manager is constantly querying the production manager as to what he or she should buy or stating that the MRP schedule is more precise than the internal schedule so that he or she should buy more and the shop make less, because purchasing is more efficient. Therefore, he or she ought to buy more and the factory do more assembling than component part making. Then we had the inevitable squabble over whose schedule is correct. So, the logical answer is, why not have the same computer-driven schedule for both: the purchase of material and the allocation of factory resources in a compatible single schedule. The logic is irrefutable and, hence, the creation of MRP II.

By the mid-1970s it was apparent that the fundamental scheduling logic of MRP could be used for capacity planning, shop floor control, and shop floor scheduling as well as for the original intent of inventory control and purchasing. Planners could use MRP-generated information to make valid, executable shop floor schedules well meshed with other manufacturing operations. This means methods, time standards, planning, and feedback measurements could all be incorporated into one grand design for controlling the manufacturing process. Oliver Wight, one of the early developers of MRP theory, is credited with inventing the concept of this unified activity. He called it MRP II, manufacturing resources planning, and was the leader in showing how a materials management scheme could be enlarged to optimally control the entire factory. He was a visionary who has since been proven to be correct.

We now know MRP II as a component of the overall CIM philosophy of integrated activities for optimizing the total output of an industrial concern. If we look at the seven steps of the Manufacturing System (see Table 2.1) we can see that MRP II makes up the majority of steps 3, 4, and 6 and small portions of steps 1, 5, and 7. Therefore, it is not surprising that if, as many contend, CIM is the automation and integration of the seven steps of the Manufacturing System, then MRP II is the entire entity of CIM. We know this to be not true; MRP II is not the entire entity of CIM, but it is easy to see how such confusion came into being. It is true that MRP II is a major component of CIM.

Let us now devote some time to describing specifically what MRP II is and does, from the original theoretical viewpoints. Keep in mind that MRP II developed before CIM became the dominant focus of business management theory. MRP II consists of

1. Capacity requirements planning
2. Order release planning
3. Operations sequencing

What these three phases do in aggregate is look at the required delivery date of the finished product and reserve for every component of that product sufficient labor, material, and production tools/processes to produce the product on time. The basic computer algorithm is to schedule all operations backward from the customer's need date, thus ensuring that all operations will start on time to meet the need. The computer does this for all the thousands of parallel and series operations required to produce products on an on-going basis, thus ensuring optimal use of all facilities. This, as we can see, is not at odds with JIT theory. Once the needs are established, we can use either a push or a pull production control philosophy to comply.

The basic data required in the MRP II database are orders and their due dates and capacity/capability information. The latter is often referred to as the manufacturing resources file and current status file. We can see how important the status of equipment and cycle times become for the success of MRP II. One of the early problems with MRP II was that it assumed infinite capacity and capability. This, of course, is unrealistic and very soon resulted in addition of limits as imposed by the manufacturing resources files. MRP II still did not become fully operational until bottom-up CAPP came into common practice. This allowed engineering time standards to validate cycle times, thus making MRP II accurate enough to operate virtually automatically.

The MRP II process starts with a non-MRP II activity, the master schedule (see Figure 8.1). The master schedule is the receipt of orders sifted through by senior management. The orders are placed in sequence by senior management with one eye on the pragmatic overall cycle time requirements (what is really possible) and the other on what promise is necessary to land the order. Often, delivery date is the sole criteria

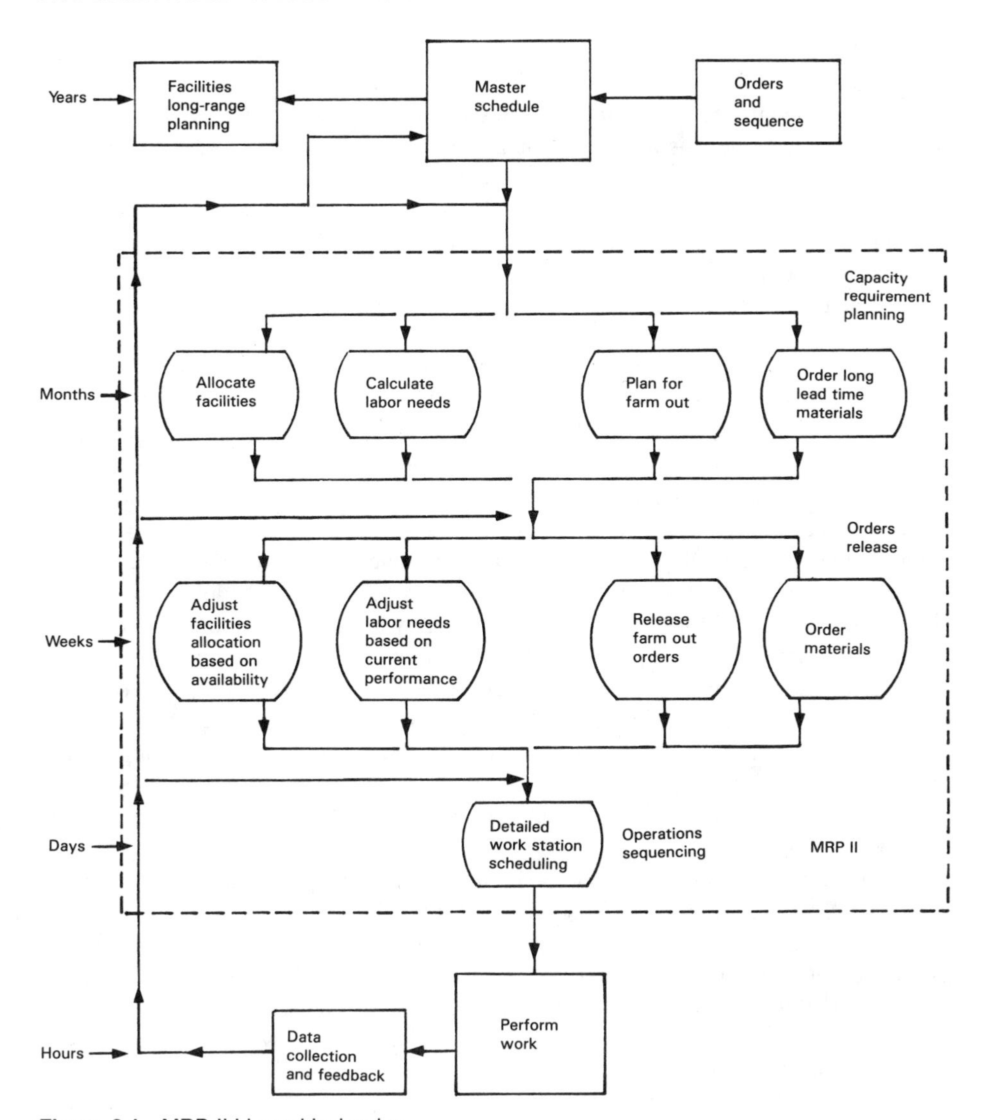

Figure 8.1 MRP II hierarchical order.

for awarding competitive contracts, so it is normal to expect that the master schedule be a tool of presidents and vice presidents to keep the company's order backlogs full. Naturally, those executives who heed the pragmatic realities of what can really be done will, in the long run, show a better success ratio than those who will promise anything to get the order. The implementation of MRP II did improve the coincidence of promise and actual performance dates.

Once the orders for the master schedule are determined, a gross overall balance is performed for loading the manufacturing facilities. Many master schedule programs run with the same, or compatible, software as the MRP II system does, so the gross balance is acceptable to the finer scheduling the MRP II system will do later. For this reason many people think that the master schedule is part of MRP II. It is not and should not be because it is a senior management planning tool and not an operations tool. The gross balance also serves another purpose. Since the event horizon of the master schedule is one to three years out, the balance or lack of balance can be used as a forecasting tool to determine if there is a need to purchase additional capacity to meet future orders growth. This becomes an equally important function of the master schedule and is usually worked on by manufacturing engineering in their role of facilities planners.

Capacity requirements planning (a part of the design/planning control leg of the CAD/CAM triad) is the first MRP II activity. This takes the dictates of the master schedule output and refines it to a six-month outlook, as compared to the one-to-three-year projection required by senior management. This activity creates the game plan the factory will follow. The output of capacity requirements planning will cascade down to orders release planning, then on to operations sequencing. Capacity requirements planning takes a detailed look at the orders to be produced and provides time allocation for the machine tools and processes by making up a sequenced schedule of what parts will go across what facilities in a given period, usually in weekly to monthly increments. By looking at demand versus capability, capacity requirements planning will also make the determination of whether enough capacity exists or certain amounts of production will require outside sources (farm out) to be completed on time. In addition, capacity requirements planning will feed shop operations the labor profile for the coming period. Finally, since capacity requirements planning takes a long-range look at the demands placed on manufacturing, it is usually the point at which long lead-time materials are ordered.

Orders release planning, part of the production and measurements control component of the CAD/CAM triad, provides detailed dates for subcontractor work, in-house work, and release of purchase orders. Its horizon time is weeks instead of the months used for capacity requirements planning. This phase of MRP II is usually referred to as the rolling thirteen week (one-quarter of a year) plan. The software associated with orders release planning gets its basic information from the downward cascade of output from the capacity requirements planning module and then fine tunes that output for immediate detailed scheduling of the next period (see Figure 8.1). To do this requires up-to-date status of factory production, which is obtained from putting current status into the database, either manually or through data collection systems.

One might ask, why this step is necessary? Why can capacity requirements planning not do this? Theoretically it can. However, the higher-order step is more strategic in nature, and, in order for orders release planning to work at any reasonable speed, data are truncated and summarized. Information detail is not required to the same degree for capacity requirements planning as for orders release planning. Certainly, if we develop our labor needs for a six-month period, we need not take into account the possibility of a flu epidemic creating fewer available work hours. However, if we are planning the factory for four weeks from now, and we know we have the flu problem to deal with, we would be negligent if we did not respond "schedulewise" to it. This is the difference between the two and also the difference between orders release planning and operations sequencing. The closer we get to actual performance of work, the more we have to take into account the vagaries of the current and near-term situations. This is the way we actually perform our management tasks; therefore, it is not unusual or surprising to find that MRP II theory has developed along those lines. Any good computer system mimics the "best ways" of doing the task manually and then enhances it with the unique attributes of the computer. In this case, the ability to do many routine calculations swiftly and accurately creates schedule algorithms.

Another reason the three major segments of MRP II are broken up is the pragmatic reality of the immense core memory a computer would need to have in order to do all functions of MRP II at one time. Since this is impractical for technology, cost, and capacity reasons at this time, the overall schema is broken down into three distinct but related software packages. Through the CIM concept of common databases all linked together, it is feasible and efficient to do this.

If we look at Figure 8.1 we see that orders release does the same thing as its hierarchical precursor. We see that facilities, labor, subcontract, and material are all considered as in capacity requirements planning, but in an action plan mode rather than in broad concept. Here we can literally smell the machine tool cutting oil and visualize individuals performing tasks. Here we are interested in the effects of engineered time standards and methods, especially improvements implemented through investigation and evaluation that will change cycle times. We are also vitally interested in materials due dates, many times down to the hour of receipt so we can get the most out of the specific work stations. Likewise, we are interested in the degree of proficiency of the various machines and processes. Are they capable of performing as the specifications say they are? The whole subject of preventive maintenance, setup, quick change, operator training, etc., comes into play in formulating plans for production. This module is the control function for the operations management team. Here they detail, plan, and track results of that plan and revise as necessary. The next module, operations sequencing, is where the commands are given to execute the plan.

Operations sequencing, part of the production and measurements control component of the CAD/CAM triad, is the front line of MRP II with respect to dispatch and performance of work. This is where production control resides in a CIM environment. This module controls queuing of work at every work station. Here the four outputs of orders release (see Figure 8.1) are used to develop individual schedules for the various work stations, usually for no more than a week, possibly two. Here we are dealing with daily scheduling. This is the very specific work instructions given to supervisors and operators on a day-by-day basis, sometimes as specific as a minute-by-minute rate.

This is where effective use of equipment and facilities is affected by proper scheduling. In this module, the goal is to load work stations to their full capacity. This is done by determining the accurate sequencing of orders so the bill of materials "goes into" strategy is optimized on the factory floor. Where required, attempts are made to meet order due dates by using reserve equipment, overtime, and farm out, which correctly implies complex software networking with purchasing, inventory control, and manufacturing engineering databases. The planning horizon is very close in (shifts and the work day), with increments being in minutes. This requires that the output be very precise to be successful. There can be nothing left to chance. The output of this module literally drives the factory. It combines all the sequences of parts making and assembly schedules and merges it with the instructions as to how those parts are to be made, how long it should take, and how correct performance is to be verified. This is by far the most complex of the three MRP II generic software modules and, as expected, the most difficult to design, implement, and execute.

It is with this module that we can take into account the pull system championed by JIT. Instead of assuming that all work is done as scheduled (thus, a push system), we can devise the sequence as requested but not issue orders to execute until verification is received that the precursor operation(s) has been completed. This, in effect, is an electronic *kanban*. This simplifies shop floor instructions by always issuing only the next order to produce. Since 1983, most modern operation sequencing modules have incorporated the JIT philosophy. We have combined strategy with tactics in a compatible manner.

As we can see in Figure 8.1, up-to-date information on shop performance to schedule is vital for a successful MRP II implementation. Unfortunately, early MRP II systems ignored this, giving the impression that all a company needed was an MRP II set of software to schedule and dispatch work. The assumption was made that infinite capacity existed to fulfill any and all demands placed on the factory. We all know this to be true virtually never, so means of supplementing capacity became necessary. These means ran the range of improving effectiveness through adoption of the pull principles advocated by JIT to subcontracting overload. Also, it became evident that an accurate feedback mechanism such as modern data collection was required to keep score as to where the company was with respect to schedule. All this was done to ensure that the database of the MRP II system remained valid.

What this indicates is that MRP II, as basic as it is to successful control of operations, is not a stand-alone system. Referring to Figure 8.1, note that while it shows the hierarchical nature of MRP II it also shows that MRP II will not work without linkage to other elements of control. In the figure we only show master schedule, orders and sequence, and data collection in addition to MRP II components, but there are more components behind the scenes, although perhaps not directly connected in a cause-and-effect manner. Those shown in the diagram are not MRP II because they serve other vital activities as well but they have to be included because MRP II is pragmatically dependent on them.

CIM is communication excellence. MRP II is communication excellence in the presentation of information for optimizing factory production. Therefore, MRP II is indeed a subset of CIM. That being the case, implementation of MRP II cannot be optimal if it is done without a well thought out CIM umbrella strategy. This umbrella strategy has sub-strategies, such as MRP II, which have to be compatible with each other and with the whole. So, implementing MRP II in a factory without giving sufficient thought, for example, to data collection or bottom-up CAPP will not sufficiently satisfy the hunger an organization has to improve its performance. Most likely, if MRP II is attempted without due consideration of the companion systems, it will not be entirely successful and will require lots of bandages to be acceptable. A taste of inadequacy will remain. With this in mind, let us isolate MRP II and see what is necessary to implement it in the typical factory (assuming there is such a thing). We will thoroughly discuss strategy for implementing CIM in its entirety (from no CIM to full CIM) in a later chapter of this text.

A discussion of the implementation philosophy of MRP II is pertinent because of the lack of a sufficient success rate. As far as I know, there are perhaps one thousand successful implementations of MRP II, by 1989, worldwide.

Contrast this with millions of firms that could use MRP II and the indicated three-to-one to ten-to-one failure rates of those who have tried to implement it. If we believe that the benefits of successful MRP II are enormous, then the disturbing failure rate of as many as ten thousand firms most likely has to do with implementation technique, not the theory itself. The fact that failure is reported so often means that the lure of having a successful implementation has not been strong enough to have those companies persevere. It would appear that not doing it right the first time leads to permanent failure. So far, few firms have been willing to give MRP II a second chance if not done correctly initially. This readily indicates that implementation is not simple, and that proper technique, followed meticulously, and with full understanding of why it is being done that way, is vital for success.

Table 8.3 is my summary (with my interpretation of the integration factors necessary to create the proper CIM environment) of the implementation steps recommended by Oliver W. Wight in his book, *The Executive's Guide To Successful MRP II,* published by Oliver Wight, Essex Junction, Vermont, 1982.

Let us now look at these steps and understand the reasons behind them. Particularly we want to understand why each step is performed and how it ties into the overall CIM philosophy.

1. *Educate all employees.* By all employees is meant all employees. That means

Table 8.3 Steps in Implementing MRP II

1. Educate *all* employees
2. Perform financial justifiation analysis
3. Select a full-time project leader
4. Integrate MRP II plans into CIM strategy
5. Get professional guidance for implementation
6. Initiate a project to perform the following functions:
   (a) Obtain accurate inventory records and maintain; BOM, 98% accurate; cycle counting and routing, 95% accurate
   (b) Structure BOM to match material moves
   (c) Check BOM concurrence with how products are made
   (d) Establish material order policies
   (e) Establish master schedule policies
   (f) Establish production plan policies
   (g) Evaluate and select software packages
   (h) Implement and debug MRP in pilot location
   (i) Implement MRP in all production locations
   (j) Close the loop, upgrade to MRP II

Note: Based on the implementation plan by Oliver W. Wight in his book *The Executive's Guide to Successful MRP II,* Oliver Wight Limited Publications, Inc., 5 Oliver Wight Drive, Essex Junction, VT 05452, (802) 878-8161 or (800) 343-0625, © 1983, used by permission.

everyone in the company from the president on down to the sweeper has to be introduced to the MRP II concept. Basically, they have to be exposed to what MRP II is, how it works, and why the company is committed to it. In addition, the employees have to be instructed on how MRP II relates to the CIM philosophy and particularly JIT.

2. *Perform a financial justification analysis.* This goes beyond the narrow focus of MRP II. Any new venture, be it a facilities expansion to addition of a product line, has to pass an economic justification hurdle. MRP II in particular and CIM in the generic mode are no exception to this dictum. By requiring an economic justification, it focuses the company on defining the ways this new technology will improve performance. In the case of MRP II, the typical performance improvements are reductions in inventory, both raw and work in process. This is usually calculable. By better scheduling we need less reserve inventory because all goods brought into the factory are pre-reserved for a particularly identified order. This is called mortgaging and is an enormous benefit to calculating precisely what is required and when. By employing JIT pull and eliminate wastes philosophies in the MRP II scheduling algorithms, we add improvements to the calculated savings of reduced inventories.

Other financial benefits derived from better scheduling include better control of the salaried and hourly work forces. We no longer have to allow for reserve personnel to be put into service on an as-needed basis. With better scheduling related to need dates, we can determine more accurately when activities are to happen and, hence, maintain closer control of the human resources and incur less expense.

Justification usually takes the form of the present net worth calculations that we would expect. The idea, like any capital investment evaluation, is to see if the company would do better by using money for something else or by simply investing it in the securities market. All companies establish hurdle rates for investments. Typically, in manufacturing industries, that hurdle rate is set around the nominal two-year return level. MRP II must meet the hurdle rates established. If it does, then it will pay off handsomely.

One caution: Do not allow finance to establish a special hurdle rate for CIM-related projects. Sometimes this is erroneously done because the finance manager feels that in evaluating these types of projects they have too many soft justifications, and, hence, we must err on the conservative side. This is nonsense. There is no more difficulty in computing saving through better inventory and management control than that of computing savings from an N/C machine making more parts per hour. The N/C machine only saves the company money if it can really use the additional parts and, hence, sell them. The reduction in inventory is real savings for every level of production output. Likewise, calculating savings through the elimination of staff positions or better use of those staff personnel through better scheduling is not a difficult task. There tends to be a bias against all new technology and often harder, more rigorous, justification is asked for. This is unfair and irrational and should be resisted. In the case of MRP II and its tactical cousin JIT, this can often be overcome by proper education of the company's decision makers.

3. *Select a full-time project leader.* Recognizing that the ratio of success to failure is not adequate, there is no doubt that a full-time manager is necessary. This full-time leader must make the effort to become conversant in the technology and also

must believe in its value to the company. He or she must become an effective champion of MRP II. The manager of the project has to be familiar with the company's policies and its culture and be capable of maneuvering through the specific political mine fields that will be erected to challenge MRP II. The successful project manager must be a senior member of management with access to the most senior decision makers. He or she must have excellent people-management skills, be well respected, and come from what is considered a main line organization. This excludes management information systems (MIS). MIS connotes a manager who is narrowly focused on the operations of computers, not on their uses. This is unacceptable. The leader of an MRP II implementation project should be the manager who will derive the most benefit from it. If this is impractical, then it should be the senior lieutenant of that person. Heading up an MRP II implementation project should be considered a significant career enhancement for the manager who successfully carries it out.

4. *Integrate MRP II plans with the overall CIM strategy.* MRP II is a component of CIM. There is no debate on this fact whatsoever. CIM derives its strengths from the common database philosophy, so the major integrating planning activity for MRP II implementation is to ensure that the databases necessary for MRP II are compatible with other CIM modules. For example, the bottom-up CAPP databases must be able to communicate freely with the MRP II modules. This means that selection of hardware and software cannot be done independently. We cannot have a relational-type database for CAPP and expect it to function effectively with a hierarchical architecture database for MRP II without some degree of conversion work at both ends. We now know that such discontinuities can be made to work together. But should we consciously choose to do so?

The MRP II implementation team has to understand how their output will fit into the overall company's CIM plan. This means they must understand all aspects of the CIM plan before embarking on MRP II implementation. In fact, it implies that MRP II cannot be successfully undertaken until there is an overall CIM plan. Many companies fail to realize this, and this may be a major contributing factor in the unacceptably high MRP II implementation failure rate.

MRP II will fail if it requires a different set of control software and hardware than the company's N/C machines, data collection systems, and other computer-driven systems. This can be avoided through properly conceived integrated planning. The confusion caused by the likely manual conversion from one system to the other will doom the entire enterprise to a tower-of-Babel–type failure. For this reason, MRP II must be fully integrated into the CIM plans before implementation commences. This does not mean CIM has to be in place. That would be nonsensical. CIM is made up of major components like MRP II, and there is no reason why MRP II cannot be the first module to be implemented. The golden rule certainly applies in this situation: "Do unto others as you would have others do unto you." This requires integration, and this is what CIM is all about.

5. *Get professional guidance for implementation.* This is good old fashioned common sense. Unless the company is blessed with a bonafide expert in MRP II implementation, it makes sense to have MRP II expert consultants help with the planning of the project. This does not mean that the consultants take over the project.

That would be tragic, because success depends largely on the various users' achieving ownership positions with the implementation. By letting consultants take over as leaders and doers, the ultimate users get intimidated and resentful, which leads to failure later on.

Consultants should be used as catalysts, not as leaders. The idea is to learn from someone else's mistakes and not make them yourself. Properly qualified consultants are conversant in the techniques for successful implementation. They have been down that path before. They are normally quite valuable in guiding the implementation team through the various mine fields once they understand, in a broad-based way, the client's particular situation. Consultants never achieve the level of company-specific knowledge the in-house management team has. That is not the nature of their expertise. They only have expert knowledge of their specific subject matter. We hire consultants so they can transfer some of that knowledge to us for our use in our particular and peculiar situations. This means that even if we wanted to, it would not work for a consultant to take over as the leader of the project. It is not fair, and you cannot expect the consultant to be as conversant with the client's situation as the senior manager placed in charge of the project.

6. *Initiate the project to carry out the implementation.* The first thing the project manager does is construct, or have done, a pert or critical path method (CPM) chart listing all the activities that have to be performed to successfully complete the project. CPM is recommended because it shows the time frame relationships between the various activities. It also shows what the prime scheduled activities are that have to be kept to the budgeted time for the entire project to be accomplished by the completion date. Once the schedule is agreed to, the following basic activities have to be accomplished. There undoubtedly will be more, but these represent the so-called generic items.

- Obtain and maintain accurate inventory records. This means know what materials are available to be worked on at any specific time. This is the domain of little MRP. Here we see the need for accuracy in inventory counts and in purchase orders outstanding. The basis of this accuracy starts with the bill of materials (BOM) and the routings that it dictates. The results of practical experience show us that the BOM has to represent at least 95% of all items needed to complete fabrication of the product. Also, routings have to be in the vicinity of 98% accuracy for the operations sequencing module to be effective.

In conducting the implementation project, the investigation of the BOM and routings accuracy is the traditional first step, for good reason. It is impossible to have any semblance of a controlling dispatch system without knowing the build sequence of the product. This is the first "know": know how to make the product.

This is the first area where I believe a fatal mistake can occur. The second "know" is often ignored. The second "know" states that knowledge as to how long it takes to make the product is equally important as the procedure for accomplishing manufacture. It is silly to think a factory can be scheduled without intelligence concerning cycle time for each operation. Somehow, this basic scheduling need was ignored by early proponents of MRP II. I do not understand why. Suffice it to say that as part of the process of verifying the accuracy of the BOM it is also necessary that the vital engineered time standards are also 95% accurate. If not, the factory will always be in a catch up or slow down condition, oscillating about an unknown mean. If these oscillations become large enough (unfortunately, they often do), then scheduled times will not represent reality or even a close enough approximation. The irony of this is if all we can guarantee is an approximate schedule, then all the work of implementing MRP II is not worth anything. Therefore, investigating the BOM and routing accuracy must also include a plan for ensuring engineered time standard accuracy.

• Structure the bill of materials to match material moves. This means make sure the BOM is compatible with the way materials are moved within the specific factory. Sometimes, a BOM is used for several plant sites, and material fixturing is not the same, so an operation may, by necessity, be reversed with a precursor or successor operation. This could mean material dimensioning may vary in the raw state, for assembly purposes, and be trimmed later on to finished conditions. This necessitates changes in cycle times, which, in turn, could affect the overall scheduling algorithms' accuracy. A similar change situation may occur if methods improvements are incorporated and the BOM is not revised. If this goes on too long, accuracy and, hence, control deteriorates.

• Check for bill of materials compatibility with the method. This is the CAPP evaluation step. Methods are devised to make and assemble the product in the most efficient way. The BOM mush reflect the optimum "goes into" pattern. If not, the MRP II system will perpetuate a less than optimum productivity performance. The goal of MRP II is to optimize manufacturing, not simply to follow a preplanned schedule with great accuracy. The ideal of CIM is highlighted here. We have excellent materials control and excellent scheduling through MRP II. In another CIM venue, we have excellent capability for creating optimum methods and corresponding engineered time standards. Through the integration philosophy of CIM, we can merge bottom-up CAPP, MRP II, data collection, etc., through the common databases and obtain the best possible production system for the given set of conditions.

• Establish material order policies. Based on relationships with vendors, we need to set minimum order quantities and lead times for orders to be delivered to the scheduled need dates. This is an excellent opportunity to introduce JIT eliminate waste principles with vendors to achieve win-win situations. In fact, incorporating these JIT inventory reduction principles with vendors must be a conscientious part of the implementation plan.

Quite often the opportunity to link JIT with MRP II is missed because the implementation team narrows the focus too much. Inventory policy is one of the drivers of schedule. It certainly affects the company's cash flow in a big way. If we need to carry more inventory because a vendor is not aware of the capabilities of saving money by adopting JIT principles, then the implementation team management is not doing its job. Just as the BOM must reflect the best production method possible, so too should the inventory schema. This is normally done by reviewing the JIT inventory related principles, understanding their implications, defining how they can be incorporated, and including plans to do so in the overall PERT or CPM network. Here we are employing a JIT tactic to achieve the strategy envisioned with MRP II use.

• Establish master schedule policies. This is not an MRP II implementation step but, rather, gaining an understanding of the constraints under which the team must operate. Master schedule, as described previously, is as much a political document as a firm procedure under which the company operates to take orders and make promises to customers. How this operates has direct impact on the capacities requirements planning module.

The implementation team will not set policy for the master schedule. Instead, they will investigate how it works and how it will affect MRP II. If its effects are acceptable, fine. The team goes on to other tasks. If, in their opinion, the procedures are detrimental to an optimal MRP II implementation, then they must explain why to senior management and make recommendations for change. The decision then lies with senior management as to whether or not they agree and will make the changes. Whatever the outcome, the implementation team must work with the results and effect as good as possible an implementation.

• Establish production plan policies. This has to do with priorities of what product components get first call for the various machine tools and processes. It is necessary to do this to avoid having bottlenecks occur needlessly. Without setting priorities, it is conceivable that more than one order may be scheduled for the same work station at the same time. This is not a simple task. It will require a detailed look at the methods of producing, the numbers of each component to be produced, and the promised due dates of each. The scheduling algorithms built into the operations sequencing module can become quite complex.

Recently, we have seen that the pull system championed by JIT affords the opportunity to minimize complexity by only scheduling assembly and setting all other operations on an as-needed rather than on an as-scheduled basis. This simplifies the computer's task by having it run the scheduling algorithms against fewer constraints. The decision tree for selecting the highest priority part for the work station to work on diminishes by orders of magnitude. In practice, what this means is that parts needed to make a customer ship date by assembly always get priority. In fact, the order to produce work at a work station is not issued until it really is needed at the succeeding work station. This is a true productivity enhancer in that parts are only made as required, and not just to satisfy a pre-ordained schedule. This makes tremendous sense. This is another example of how the JIT tactical approach makes the MRP II strategy work better.

• Evaluate and select software packages. This is the second area where a fatal mistake can occur. It is the implementation team's obligation to make sure that the software chosen to run the MRP II program is entirely compatible with the company's way of doing business. This means that not only does the software have to do all the functional requirements deemed necessary by the team to run its view of what MRP II should be but, more importantly, that it do so in a way that is comfortable to the entire organization. In practice, this means that as few new data requirements and concepts as possible be introduced. Every change introduced creates another opportunity for failure. Therefore, changes must be kept to a minimum.

Most successful implementers have chosen to keep as many pre-MRP II procedures in place as possible as long as those procedures are not contrary to MRP II principles. The same terminology is kept as far as is practical. If a routing card in pre-MRP II is called a move ticket, it should continue to be called that after installation. The strategy must be to ensure that the work force has familiar pillars to hang on to.

This desire to make change as minimal as possible means that the software selected should be adaptable to the company's way of doing things. All too often, software selection is left to computer systems personnel who are not familiar with the internal procedures of manufacturing. When this happens, the choice is often further delegated to the software vendor working with the internal systems personnel. The abrogation occurs because the internal systems staff often feel that the software producers know more about MRP II; therefore, they are also knowledgeable as to what will work in the client's factory. The internal systems people certainly will be susceptible to anything the vendors tell them if they are not very close to the manufacturing process. This scenario could and all too often does lead to unmitigated disaster. We find that the software is so alien to the firm's way of doing business that major shifts in procedure are necessary to accommodate the software. When this happens, the path to failure becomes a six-lane super highway and only herculean efforts will avert failure.

How do we minimize the chances of something like this happening? Very simple. Make the users responsible for selecting the software, not the technicians who are responsible for keeping the computers running. Also, make sure that the users understand what MRP II is supposed to do. When we combine this knowledge with the detailed understanding of how the factory really works, we have a very good chance of selecting software that is acceptable to the company. This software will probably work, and, since the users are involved in its selection, they have ownership in it and will work hard to make it usable.

Keep in mind that a vendor's prime responsibility is to sell his or her product. In the vendor's mind, selling the product is always prime. The fact that it may not work as well as perceived by the user is a detail that can be handled, or so the vendor rationalizes. Software vendors, in particular, have become adept at placating clients, to the point of getting clients to accept total responsibility for failure. I cannot begin to count the number of times I have come across unhappy computer system users only to discover that they are practically "in love" with their software vendors. This, even though the vendor knows that he or she sold the client something beyond the scope of his or her needs or understanding. Vendors rationalize this by stating, "If only the client would do everything I told them to do, the software would perform perfectly." Unfortunately, the vendor does not understand or want to understand the true nature of the client's need, nor will he or she ever be in the position to do so. So the vendor rationalizes failure as the client's fault, and the client, not knowing any better (since the computer systems people probably bought the software), agrees and absolves the vendor from responsibility. The chances of this sad scenario being played out are greatly diminished if true end users buy the software and they recognize that the vendor's loyalty has to lie in making a sale at all costs. When everyone understands the other's prime driving forces, then proper software transactions will occur.

A more complete and descriptive procedure on how to implement CIM software is presented in a later chapter.

• Implement and debug MRP in a pilot location. There is no typographical error here. I mean to specify little MRP. It is important to achieve a small-scale success before we tackle more complex and dynamic situations. Materials requirement planning means limiting our focus to obtaining materials in accordance with the production schedule and not, at the moment, concerning ourselves with the entire seven steps of the Manufacturing System. We focus on ordering materials, their receipt and entry into the inventory control database, and relieving that database as the material is dispatched to the work station. In practice, this means creating vendor files, for automatic order placing and tracking, and receiving files, to collect data when materials arrive. It also means putting into operation algorithms for generating purchase orders when certain key points are reached, such as minimum stock levels. This requires organizing the stockrooms such that the logic used for recording material inputs and outputs is consistent with good practices and is still compatible with what the company feels comfortable with.

We chose a pilot operation because materials control most often involves large numbers of different types of parts and commodities. We want to learn how to live with the software and procedures on a scale small enough that if mistakes are made (and they will be) then they are not large enough to seriously affect the entire company's performance. We want a laboratory to practice with, where we can debug the software and learn and become competent in its use.

After MRP is fully debugged and in daily use in the selected pilot area, we can consider expansion into other areas. The strategy is to repeat the procedure used in the pilot area in other areas until MRP is up and running throughout the entire company.

• Upgrade to MRP II. This is the final implementation step. Training is completed, and MRP is running successfully throughout the company. Now we include the people and facilities resources utilization scheduling with materials scheduling. Once more we use the pilot area implementation strategy for the same reasons as were just mentioned, only now the stakes are higher. We are attempting to control the dynamic nature of manufacturing by means of the input and output of information concerning the status of multiple and different but concurrent activities. We are reaching the point where production management has a knowledge of what is happening in the factory, right now! The goal of the industrial revolution, to control the seven steps of the Manufacturing System, is now possible. This will happen if the implementation steps are adhered to and care is taken to understand the ramifications of all the action steps described.

There is a need now, before concluding this chapter, to tie MRP II and JIT together in a manner that illustrates they are not competing technologies but a strategy and a tactic that optimize productivity. The fact that one is a strategy and the other a tactic means that they cannot compete at all, and to think of them as equal teams on the playing fields of manufacturing is just plain wrong. That would be the same as thinking that automobiles and engines are equivalent. An engine is a major component of an automobile. So, too, JIT should be thought of as a major component of MRP II.

I know this is not what one would deduce from reading the various articles and books about the two. But logic, driven by common sense, dictates it to be true. We have seen that the philosophy of JIT, as well intentioned as it is, cannot work for long without some significant assist from the CIM concept, and the major CIM component that is associated with scheduling efficiency is MRP II. Hence, JIT is a means of organizing MRP II to be more efficient. This makes JIT a tactic supporting the CIM scheduling strategy, called MRP II. MRP II can exist without JIT. It just would not be as effective. JIT cannot exist without a grand scheduling design (e.g., MRP II).

Figure 8.2 shows in generic summary form the data flow of MRP II. Note that MRP II sends work to work stations in a hierarchical manner. It does not, in the traditional mode, concern itself with whether or not the work station (where the work is to be performed) is ready to accept the job. If the schedule demands work to be dispatched, then it does. The operations sequencing module simply does its duty and dispatches.

We rely on data collection input into higher hierarchical modes to change instructions to the operational sequencing mode, if major modifications to plans are necessary. Therefore, we can conceivably dispatch work even though there is still incomplete work at the work station. This simply creates a queue at the work station, a backlog. Since backlogs have always been accepted, then, within bounds, MRP II also accepts backlogs. MRP II is immensely more powerful and productive than pre-MRP II scheduling because it operates in a dynamic mode, not a static one. This means it can change schedules at will if we have major traffic jams. This is something that pre-MRP II scheduling could never do without enormous effort and with any guarantee of success. The question becomes, what is a production traffic jam? How much will we tolerate? Traditionally, we have tolerated more than we should, probably because in pre-MRP II scheduling there was very little that could be done.

Or, at least, that was the traditional thought. Nothing could be done to eliminate traffic jams. We were decidedly wrong. If we had paid attention to the dictates of good industrial engineering practice, we would have known a long time ago that traffic jams did not have to be, and even a traffic jam of one excess job at a work station need not be tolerated. We now know this as the eliminate waste concept of JIT. So, JIT tactics come into being that show, through *kanbans* for simple production control systems, that there need not be more than one job at a work station at any one time. This means our grand scheduling strategy algorithm, MRP II, should be modified.

Figure 8.3 shows what happens when the tactic of only one job at a time allowed

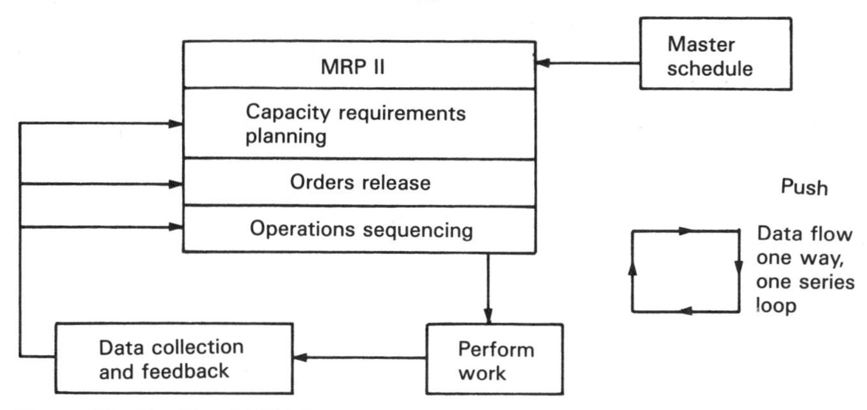

Figure 8.2 Traditional MRP II.

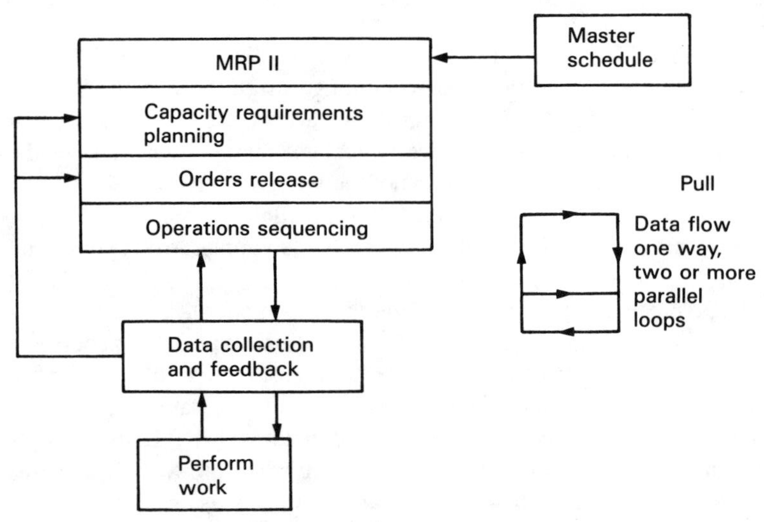

Figure 8.3 Combined MRP II and JIT.

at the work station is added to the MRP II generic flow. We simply place data collection between the operations sequencing module and the work station. Thus, the work station can talk directly to the scheduler and state when it is ready to receive the next job to be worked on. We have made the whole MRP II system more responsive, more sensitive to real world conditions. This makes us even more efficient because we will have a more precise control of when material will be needed. This cascades all the way back to the supplier who in turn saves resources by directing them only to true need. Now, this improvement cannot be practical without a grand strategic scheduler, MRP II. However, it would not even be considered if the tactic of eliminate waste had not been espoused in the first place. A simplified way of doing it, the *kanban* tactic, had to be demonstrated before the CIM scheduling strategy could be modified.

We see that the push system shown in Figure 8.2 is an adequate but not optimum way of achieving dynamic scheduling. When we combine JIT principles with MRP II strategy, as shown in Figure 8.3, it becomes evident that the pull system is superior to the push system. This leads to a better dynamic scheduling module within the CIM concept.

This is a simple example demonstrating the compatibility of the JIT tactics with the MRP II strategy. It points out that these philosophies of operations are mutually compatible, and the prudent manager must consider using both simultaneously to achieve optimum results. To think that JIT or MRP II can be used independently, ignoring the existence of the other, provides a shrewd competitor an obvious path to a competitive advantage. Competent managers simply do not allow this sort of thing to happen.

Chapter 9

Group Technology: The Indexing System of CIM

In this chapter we will explore the uses of group technology (GT) in the CIM environment. Sometimes, GT is referred to as the RNA/DNA of CIM. This reference to the genetic code is a suitable synonym. As the genetic code describes the biological entity, so GT describes the physical entity we manufacture, along with the process employed. RNA/DNA describes in precise code what a biological system is; GT does virtually the same thing for a manufacturing system. GT has uses in design, production control, planning, and factory layout. It also has found its way into marketing strategy planning. We will look at the various manifestations of GT use in the CIM environment and show how it can be instrumental in making CIM a practical reality. It is this precise coding capacity that allows us to optimally use the power of common databases to provide easily accessed integrated information to the many functions of a business entity. Without GT, we would find the system exceedingly complex, perhaps to the point that to make it work far exceeds our patience and comprehension.

Let us start by looking at GT as an integrating function. Since the most important factor of CIM is communication excellence, it stands to reason that some form of common language is necessary. Without a precise vehicle for communications, we stand threatened by the tower of Babel syndrome (e.g., talking at each other, not to each other), the result being misunderstanding, too often of gargantuan proportions. This common language must have the following attributes:

1. Precise meaning: No double, triple, or whatever, definitions for the same phrases.

2. Tightly structured: No meandering long expositions to get to the meaning of the phrase.

3. Easy to use: Quick and convenient, easy to learn.

Also, the common language must be of use across the entire business function. All functions must want to use it because it has benefits for the users. It must also be multi-function beneficial. This means it cannot be beneficial only to the user but to others as well, although not necessarily in the same way. Another way of saying this is that the initiator (the speaker) gains value as well as the receiver (the listener). For example, we cannot expect design engineering to put GT codes on drawings they produce if there is no direct benefit for them. Typically, the benefit to design engineering is easy recall of parts previously designed that may be similar to or the same as the current need. Manufacturing engineering, in this case playing receiver to design engineering's role of initiator, gains value because it can use the sorting the code allows to determine the proper method of producing the part. The manufacturing engineer, by means of the code, can determine if similar or identical planning exists, thus simplifying the performance of his or her task.

We are interested in communication excellence because we want to minimize the non-value–added time of the manufacturing cycle. This is done by processing information quickly and accurately. We need to ensure that the receiver is getting all of the message (100%), not just a large portion of it (98%). Not having 100% transmission accuracy capability by means of plain language requires establishment of a feedback check network. This feedback is not for job status accomplishment, but for information receipt accomplishment.

"Do you understand what you have to do?"
"Yes."
"Then tell me what it is so I can see if it agrees with what I said it is."

GT affords the opportunity to transmit information with only one meaning and that can be easily verified. This goes a long way toward achieving communication excellence. Use of a precise code ensures that all involved are talking about the same thing, and there is no ambiguity concerning the meaning of the communication. This means that the feedback check network can be eliminated by means of a database using GT code as the prime data transmission source.

Figure 9.1 demonstrates what the goal of GT is with regard to information transmission accuracy. Observe that the section of the figure called "common language" could be the flow of the spoken word of a supervisor to an operator. Section *a* is the supervisor telling the operator what he or she is to do. Many times it ends right there, and we suffer the consequences if 95% receipt of information is not sufficient to get the job done properly. Perhaps this is where the phrase, "5% never get the word, and they kill you," comes from. Section *b* represents a diligent supervisor trying to determine if the operator really understands what he or she was told to do. Sections *c* through *d–n*th represent the infinite number of iterations that may be required for enough of the message to be transmitted to ensure success. This, of course, can be very subjective.

Common Language

(*a*)

transmitter (100%) — send → receiver (95%)

(*b*)

transmitter (100% − 95% − x%) ← confirm — receiver (100% − 95%)

(*c*)

transmitter (5% + x% + y%) — send → receiver (5% + x% + y% − z%)

(*d*–*n*th)

transmitter ← confirm — receiver

(iterate until needed accuracy level achieved)

Group Technology Code

(*a*)

transmitter (100%) — send → receiver (100%)

(*b*)

transmitter (100%) ← confirm — receiver (100%)

Figure 9.1 Information transmission accuracy.

Contrast this with the section of Figure 9.1 called "group technology code." Here we see 100% transmission and 100% confirmation because the code can be precise in its meaning. It is 100% precise meaning that is required for an optimal business entity, which is now possible by means of GT integrated within the CIM common database. CIM requires communication excellence. With GT coding, we can take quantum steps in achieving that aim.

With communication excellence, we can achieve more efficient job shops by using the Principle of Sameness, on which GT is based. By employing the Principle of Sameness, we can emulate flow manufacturing while making dissimilar end use parts. Communication excellence allows us to cut through the clutter of information and accurately select commonalities that are useful. We say that GT enhances our capabilities to achieve communication excellence. Let us take a look now at the theory that provides the foundation for GT. We do this so we can understand how it results in improved capability by means of better manipulation of information.

To emulate flow manufacturing, we have to approach a condition where all parts crossing a work station, or a series of work stations, are very similar in the processes required to make them. Flow manufacturing assumes identical steps in making the parts (e.g., the parts are all the same in all aspects). To emulate this in a job shop, we have to have many common characteristics for all the parts going across the designated work stations. The more commonality found, the closer we can emulate flow manufacturing; and the closer we are, the more efficient we will be. Common characteristics, such as same basic shape or same basic machining requirements or the same basic assembly techniques, are the types of things we have to look for if we are to sort parts and be successful in emulating the flow manufacturing model.

It is the Principle of Sameness that establishes the concept we need. The Principle of Sameness tells us that parts having similar shapes, materials, and manufacturing requirements can be produced in a similar manner. It is the basic theoretical concept of GT. The Principle of Sameness states that parts can only be grouped on the basis of similar geometry, processing requirements, or end uses. Any other grouping has no meaning in an industrial sense. For example, if we group by cost to produce, then it has no industrial meaning because it does not define a commonality showing similar characteristic information that is of interest in making the part or in its end use. It is only by chance that the costs are the same. Conversely, if we group by material, then we limit the number of manufacturing techniques that are possible to make the part and also limit the design variables attributed to the materials selection. Grouping by materials is a subset of either geometry or processing requirements. This means that GT is interested in looking for all and any types of similarities between many diverse and varied parts as long as those similarities make industrial sense. Once these similarities are found, we then look for manufacturing or engineering advantages that can be had from such groupings; thus the title, group technology.

The fact that we are sorting by similarities leads us to consider the use of computers to help in that sorting. CIM collects huge amounts of data and disseminates that data to many diverse users. It becomes evident that a technique is necessary to make this collection and dissemination as efficient as possible. This is where GT comes into the picture. We can write computer programs using GT codes to classify data. Since CIM is about efficient operation of manufacturing businesses, it stands to reason that there ought to be enormous benefits to be had by grouping data in ways that are industrially significant, as espoused by the Principle of Sameness. This, as we can see, is true, and we end up with a multitude of uses for GT. Using CAPP, as an example, we can see why the use of GT is required if we are to classify plans for easy recall. We need to do this to make the database manageable.

One of the ways of finding sameness through GT is by means of classification codes. This is one of the most powerful uses of GT and one of prime importance for CIM. The next section of this chapter will explore these codes.

We will now investigate classification coding systems and their uses. Classification and coding is a way of exploiting the Principle of Sameness. As was mentioned previously, the Principle of Sameness means that manufactured parts can only be grouped by geometry, manufacturing processes employed to make them, or by end use of the parts if groupings are to be significant in an industrial setting. This also implies that the logical correctness can be extended to assembly of parts as well. By using the thought that the groupings have to be significant for industry and expanding on that, we can assume that there would only be a finite number of operations that can be performed on any set of raw materials. This means that the entire universe of possibilities does not have to be considered in developing a classification and coding system. It means that the number of codes will not be so large as to make them impractical to use. Instead of having to deal with trillions of possible combinations, we bring it down to only tens or hundreds of millions at most. Thus, what initially would have been thought to be an endless variety of manufacturing operations and geometric shapes can be severely pruned to a manageable amount.

This is theory. Let us see if it passes the engineer's eyeball test and if the typical factory can be adequately covered with a practical code (e.g., one that is feasible to use with a reasonable size computer). Let us take a material, such as steel. What can be done with steel? We can machine it by many different processes, such as drill, ream, bore, mill, plane, grind. We can join it by welding, soldering, riveting, bolting, interference fitting, etc. We can also shape it by forging, forming, bending, and sawing. We can change its physical and chemical properties by heat treating, stress relieving, melting down, and casting with other elements. There are probably other things that can be done. But the fact is that these are common operations that can be accomplished with steel. Now, how many combinations are there for all those things mentioned? Many for sure. I have listed eighteen, and each of these eighteen could have many variations. For illustrative purposes, let us say that each has five common variations, and it is possible to equate all possibilities with all combinations of the eighteen with the five variations (which in practice is not possible; for example, you cannot forge and weld at the same time). This means the number of combinations possible would be eighteen raised to the fifth power; or, that a code system would have to be able to distinctly categorize 1,889,568 different combinations. Say our factory works with five different categories of steel and three different categories of aluminum and six different categories of brass. Then we would have five times the number of combinations for steel, plus three times the number of combinations for aluminum, plus six times the number of combinations for brass. So, now our code system has to account for 9,447,840 plus 5,668,704 plus 11,337,408 or 26,453,952 combinations.

Now, does this pass the eyeball test? Can a code hold that much information? Certainly. In Connecticut we have automobile license plates with six space holders; the first three with numbers and the last three with letters. That means the numbers account for 1000 combinations and the letters account for 26 raised to the third power. Multiplying the two together gives Connecticut 17,576,000 license plate combinations. In New York they use four numbers instead of three, resulting in 175,760,000 license plate combinations. Therefore, we know our factory's number of combinations is practical for a code and for computer manipulation. I assure you that the states of Connecticut and New York do not use supercomputers for their motor vehicle registration files.

Our code, then, would have eighteen spaces for operations and one for materials. We would also want space for geometric parameters and end uses of the part. Perhaps we would also want space for assembly considerations and specifics about the number of holes drilled into the main body and other peculiarities in which we might be interested. We can add any pertinent information we want as long as it is industrially significant. How many spaces do we end up with? There is no set number. Some codes have thirty-two specific spaces. Others may only have ten. The significance is that we can create a finite model that has more than enough space to handle all the manufacturing and design information that a good size firm would ever need. It means that we can use a code to simplify the data sorting of the CIM database. We use the code to make sorting and manipulating easy, much like RNA/DNA instructs genes how to line up in each cell of each species.

Classification coding came of age with the maturing of computer use in industry. Without computers, the number of code spaces cannot be much more than four or five. Any more than that becomes impractical for manual manipulation. Five spaces allows for 100,000 combinations for a straight numeric code. If we combine letters with numbers (an alphanumeric code), the English version would have five raised to the thirty-sixth power number of combinations. This is way too much information for manual encoding and decoding. We would be dealing with encyclopedia-size code reference books, and even those might not be of sufficient size to handle septillion level combinations. So, if we need to size a code to cover a factory as just discussed, then it is obvious that a manual coding system is not going to be sufficient for any but the most simplified set of information. A data set with a potential for 26.5 million combinations is beyond the capability of even the most determined management team.

Accepting the fact that a classification and coding system must be manipulated by a computer database, we should now define the main purpose of employing the code. I have implied that the code makes management of the enterprise simpler. But, by what mechanism? The mechanism is the identification of families of parts. If we can identify families of parts, then we can employ the Principle of Sameness to group those parts by manufacturing and engineering functions. This means we can find all sorts of parts that, even though they may have different end uses, have either common geometry or common manufacturing process steps. This, in turn, means we can batch manufacture to our advantage, giving longer production runs with similar setups. We can approach the efficiency of flow manufacturing. Also, we can find the common denominator equipment(s) used in producing these families of parts, thus establishing manufacturing cells.

The establishment of cells gives job shops the opportunity to set up fully utilized production lines to make the associated families of parts. The cell, having a once through production sequence, mimics the efficiency of flow manufacture. It also allows management to minimize distance traveled by the materials and, hence, reduce WIP.

Finding families of parts is the means by which we use the code. How does the code do this? Recall that we said the code is a precise language with only one meaning for each bit or place in the code sequence. The technique is to query the user through specific questions. These questions are the same for encoder and decoder. Figure 9.2 is a simplified example of a code for valve stems. Note that each column represents a specific piece of information that is difficult to misinterpret. This means it is very precise, with no double meanings. The first digit is family type. Since it is the letter A, there can be as many as thirty-six types of valve stems possible. By starting off with family type we automatically truncate the possibilities. The second digit tells us the material the part is made of. This further truncates the processing possibilities. Whether it is steel or bronze will point us in the proper direction for process equipment capabilities. Digits three through six specify the geometric envelope that describes what the part looks like. This is probably all that is necessary. However, in addition, this simple code provides for special features that further define the part. So we see that this simple code defines quite adequately everything we need to know about what we would be

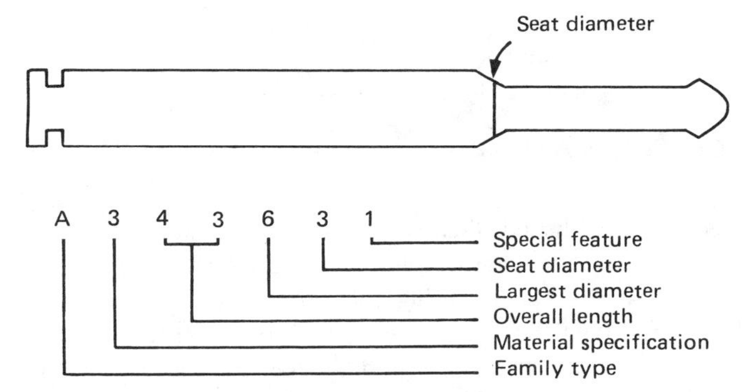

Figure 9.2 Simple GT code. (From Daniel T. Koenig, *Manufacturing Engineering: Principles for Optimization,* Hemisphere, Washington, D.C., 1987.)

asked to make. The code building procedure with the computer could be something such as the one illustrated below. This is admittedly simple, but illustrative of the process.

Computer: Select family type.
A. Steam
B. Water
C. Gas
Engineer: A. Steam.
Computer: Select material specification.
1. Aluminum
2. Bronze
3. Stainless steel
4. Steel
Engineer: 3. Stainless steel.
Computer: Select overall length to the nearest inch.
Engineer: Forty-three.
Computer: Select largest diameter to the nearest inch.
Engineer: Six.
Computer: Specify seat diameter in inches.
Engineer: Three.
Computer: Identify special feature.
0. None
1. Groove to hold actuator lever
2. Drilled hole(s) above seat
3. Drilled hole(s) below seat
4. Corrosion resistant inlay at seat diameter
Engineer: 1.
Computer: The code you have selected is A343631.
A. Steam
3. Stainless steel
43. 43 inch length
6. 6 inch largest diameter

3. 3 inch seat diameter
1. groove to hold actuator lever
Do you agree with the selection? Yes. No.

Engineer: Yes.
Computer: Thank you. Code is A343631.

Now that the code has been established, then we can query the database to find out if there are any matches for what we want to find. Suppose we are interested in finding similar parts for establishing a manufacturing cell. In this case we would be interested in searching for material type. So, first we search the database for all valve stems made of stainless steel. This states that high strength steel cutting tools would be needed, which, in turn, specifies high speed lathes. It also shows that no seat inlay is required because the entire body is made of a corrosion-resistant material. This tells us that no welding or heat-treating equipment will be required. Next we will search on size. Here we would probably ask the computer to search about a range that best fits the machines the company possesses. For our example let us say we have lathes that can handle twenty-inch to sixty-inch lengths, with diameters from one inch to eight inches. The company also has saws that can cut stainless steel bar stock up to eight inches diameter. So, now the computer will find all valve stems that fit the designated code profile:

Digit 1	A, B, or C
Digit 2	3
Digits 3 and 4	20 to 60
Digit 5	1 to 8
Digit 6	1 to 8
Digit 7	0 to 3

All codes that fall into this profile will be stainless steel valve stems that can be machined on the company's available equipment. This equipment will only have to be assembled in proximity to create a manufacturing cell with a defined family of parts.

This is a simple example, but it shows the synergism created between the computer and the engineer (or any other user) by means of a GT code.

Classification code structures come in two basic formats: hierarchical classification and chain-type classification. Both accomplish the same thing as far as the user is concerned. The distinction is wholly within the structure for use with computer databases.

The chain type is the older of the two and is the one developed in the 1930s for use as a manual code. In the chain type, each digit's specific location is fixed for a particular meaning. The valve stem code described previously is an example of a chain-type code. As we discovered, the first digit is always the family type, while the second digit is reserved for material type and so forth throughout the code structure. If we want to add more information, then we simply extend the number of digits in the code. This was not practical before the merging of code with computer software, but now it is not difficult at all. The advantage of this type of code is that it is easy to learn. Each space means a certain range of characteristics, and each position refers to a specific attribute. This type of code structure is satisfactory for describing the part's intended use or the manufacturing process for making it. The sole disadvantage is that the more information desired to be contained in the code, the longer the code becomes. This at times can pose problems in establishing the computer database software to handle the code. It may also require large capacity computers to handle all the data. Chain code is not as efficient a computer solution as the hierarchical code structure. This has been important up until now but probably will not be in the future when even greater computational power is available.

In the hierarchical code structure, each code character depends on the preceding one. This is a tree-type structure where the answer to each query by the computer sends the person doing the coding to a specific branch. So, each code slot depends on the previous entry, which in turn qualifies which table will be entered for the succeeding code character. This is much like the strategy used for relational databases, where the tables searched are a result of the comparison the software is being asked to make.

Figure 9.3 is a transformation of the chain-type structure of the valve stem example to a hierarchical code. We can see that the valve stem is only one of many parts that can be entered for this code. Once "valve stem" is selected, the branch relevant to that part is entered. We can see this as we go from part description to geometry description. It is obvious that the tables for flat and sphere do not apply. They are shown here for illustrative purposes to indicate that there can be many more geometry descriptions than that applicable to valve stems. The rest of the figure is specific to valve stems. However, we must keep in mind that there is much more to this code than what is presented in the figure. This is similar to the popular Lotus 1,2,3 program, where only a portion of all the spread-sheet cells are available for viewing at any one time. The material selection component is also left out to make the figure easier to study. For those who are curious as to where this would be entered, it would probably be before the part description column.

The advantage of the hierarchical code structure is a much shorter code, usually four to twelve positions. Contrast this with thirty-two or more positions for chain-type codes. Because it is more concise, we can represent a greater variety of parts and assemblies with ease and not overburden the database. The major disadvantage is the difficulty of setting up the code. To do so requires expertise in database structures and a knowledge of the parts and processes the code is to support. The hierarchical format is much more abstract than the chain type. Therefore, it is much harder to conceptualize whether the code is correct or not. The eyeball test is not very effective, so it is difficult to learn as compared to the chain type. In fact, learning the code is not the intent of those who advocate it, who state that it is an interactive database and the only thing necessary is to know how to manipulate the database. The computer does all the sorting and matching.

The style of code used is a choice based on a company's need. If the code is to support various activities in a full CIM environment, then perhaps the hierarchical type would be favored. Hierarchical gets the nod because it is a much more efficient user of database storage capability and would probably be faster. If speed were important, then chain types would not be recommended. Searching through thirty-two to forty

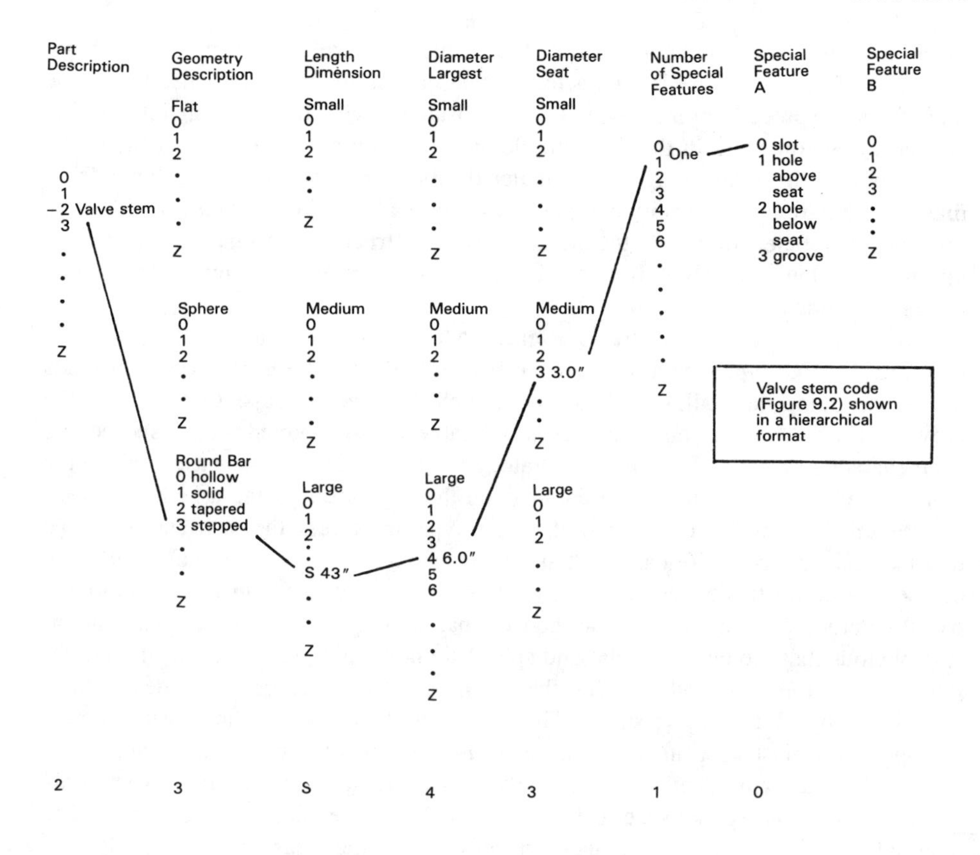

Figure 9.3 Hierarchical code structure.

digits would require a larger capacity computer than the four to twelve digits of the hierarchical version. Both types are computer interactive and friendly, so this aspect of selection no longer has merit. It all comes down to personal preference. The important thing is to recognize that the code is a significant part of a CIM strategy and not to think of it as a nice thing to have but not particularly necessary.

One of the questions often asked is how an organization works with a GT code. We have already described how the engineer interacts with the code database to code parts and how the code can be used for finding similar or exact matches for grouping activities. It may also be necessary to describe the entire design and planning process with the focus on what the GT steps are. Let us look at Table 9.1 and briefly describe the GT activities involved.

The table represents the design and planning activities most companies go through when a new order is received. It is in chronological order from the top. The first four steps are before GT is introduced as a procedure. The fifth step is the first GT activity.

Table 9.1 GT Steps in the Production Cycle

	• Design engineering defines product
	• Drafting executes the layout
	• Design engineering evaluates the layout
	• Layout reviewed by design engineering and drafting to determine number and types of parts and assembly drawings required
GT step	• Component parts and assemblies are coded
GT step	• Codes entered into the database, which is searched for "similar to" and "exact match" drawings and plannings
GT step	• If "exact match" are found, data sent to MRP II for insertion into schedules
GT step	• If "similar to" are found, design and manufacturing engineering evaluate whether they can be used if modified
	• If there is no "exact match" and no "similar to," design and manufacturing engineering start from scratch

We are asking for additional work to be performed, and this is the activity described in the valve stem example.

The next step, looking for matches, is a computer database activity and is similar to that described with the valve stem work cell example. If exact matches are found, then the work to introduce this new shop order is over. The data are simply electronically transferred to the MRP II database for action. If exact matches are not found but almost matches are, then the work scope has been greatly simplified. The manufacturing and design engineers need only evaluate how close to the current need the almost matches are and determine if it is feasible to modify one for their current need. If current need modifications to previously used designs and plannings are acceptable, then this is done, and the work is over. We proceed from there and make a duplicate. If this is not possible, then the work could also be significantly shortened because a major part of the previously made part may be usable for the current project, thus only requiring new work for a small portion. This process means that full-scale engineering work is only necessary if no matches are found because these are truly new experiences for the entire organization. Through GT and the CIM database we prevent the firm from constantly reinventing the wheel.

There are other uses of GT that have not been discussed. One very important use is to create a logical shop layout by means of the determination of family of parts manufacturing cells. This is covered extensively in my book, *Manufacturing Engineering: Principles for Optimization*. Another use of GT is with what is called generative planning, the ability to have the computer create the best planning for given circumstances. This includes the use of artificial intelligence techniques. Since this is such a new and controversial subject, I have chosen to devote a separate chapter to artificial intelligence as applied to manufacturing. In that chapter, we will introduce the concept of generative planning.

This concludes the discussion of GT as it relates to CIM. We can see how GT simplifies the CIM database sorting, transmitting, and receiving problems. In fact, CIM as a useful communication and integration philosophy dedicated to business optimization is impractical without it.

Chapter 10

Artificial Intelligence Applications in CIM

CIM, as we know, is communication excellence applied to optimizing business performance. We recognize that if we can optimize the performance of the seven steps of the Manufacturing System, then we can become competitive with any other firm, worldwide. So, with computers we strive to smooth out the bumps and wrinkles of managing our enterprises (e.g., carrying out the day-to-day doing of the seven steps). We have seen that applying computers to facilitate the communication flow gives us a distinct advantage in managing and using information when compared with the manual counterpart.

The next logical step is to have the computer suggest courses of action based on the received data and state-of-plan execution. This implies that we should spend our time on strategies and evaluating possible competitive scenarios instead of handling day-to-day tactical situations. We should let our tools do that. Just as our ancestors dreamed of off-loading the physical discomfort and stress of manual labor first to animals and later to machines, we now dream of off-loading the more pedantic mental labors to quasi-thinking machines. We see with computers the feasibility of doing just that and, to a certain degree, we have already succeeded.

So far we have explored using computers to do essentially clerical and by-rote activities, and we have found that these practices are highly beneficial, sometimes astonishingly beneficial. Now we are ready to look one step farther. Is it possible to have the computer actually make decisions for us, to actually do the tactical thinking work? The answer is a definite maybe. It lies in the field of artificial intelligence (AI), which shows much promise. In this chapter we will explore this possibility and try to understand the ramifications for the factory of the future.

Artificial intelligence has to do with creating thinking machines that emulate the human thought process. It is an academic pursuit largely confined to research universities and, hence, is remote from the day-to-day drama of running an industrial enterprise. Researchers, mainly psychologists and some electrical engineers, spend much of their time deciphering how we mentally process information so that they can model the patterns and try to create computers and computer programs that can do as we do. They are interested in plain language programs, talking computers, intelligent robots. It sounds like science fiction, and perhaps some of it is. One useful result, however, has emerged from the many decades of research. It is what we call "expert systems" or "knowledge systems." Researchers have learned in a general sense how to emulate the process a human expert goes through in solving quasi-abstract problems. This has great potential for making expert assistance available to many far-flung organizations. This is analogous to the benefits of N/C machines over manual machines because the N/C machine always emulates the best operators. We will discuss expert systems in more detail later.

The state-of-the-art for AI today is that no computers think. Computers do not have the capability for creative thought or initiative. All computers are dumb machines that doggedly follow arithmetic and logic instructions. This means they remain superfast operators for addition and subtraction and for if/then logic statements. The breakthrough in AI, expert systems, means that programming has gone beyond the simple if/then logic. In practice, it means if/then has evolved into setting up logical inference sequences and the ability to continuously add new information to make it easier to infer and formulate solutions. Even with this development, computers do not think nor create new ideas. AI, for practical business application, then, is an ability to go from direct cause-and-effect programming to a cause and most plausible effect programming. We will explore the state of the art of expert systems and then see how it can be applied in an industrial arena. We will then look specifically at aspects of CAPP and SQC and see how they can be enhanced with expert systems.

Expert systems are now commercially available; however, there is still some doubt as to their viability. They appear at this time to be satisfactory for limited applications. "Mycin," developed by Stanford University, by most reports, does an adequate job of helping medical doctors diagnose blood disorders and meningitis. General Electric has developed a maintenance trouble-shooting system for railroad engines, which was produced by their Locomotive division. Another successful example is the General Motors Charlie Program. This was developed to assist factory maintenance personnel diagnose machine failure causes in their automobile factories. I am sure there are many other successes that have not reached the popular press. What this points out is that the technology is still in its infancy and by no means a mature application of computers in industry as, say, N/C technology. As I have said, the successes are all for limited applications. We shall see why shortly. This does not mean that some day expert systems may apply across the board to an entire CIM database and all its varied uses, but now it does not and probably will not for a long time.

Let us look at the generic structure of expert systems and investigate how they work.

Figure 10.1 from *Expert Systems: Artificial Intelligence in Business*, by Paul Harmon and David King, published by John Wiley shows an overview schematic of an expert system. I have added the brackets "application specific" and "generic shell" to show the major divisions of the software and to help explain the various components.

The commercially available portion of the software is the generic shell. This contains the software that manipulates the information in the applications-specific section (known as the knowledge base) supplied by the user. The generic shell, also called the inference engine, contains the if/then probable then logic that allows the user to obtain the "expert" answers for which he or she is looking. Let us define block by block what happens in the organizing and using of an expert system.

Working Memory

The working memory is not really part of the finished expert system. It is the memoirs and recollections of the expert(s) who is (are) being debriefed to form the knowledge base. We introduce a new job description to do this activity. It is called a knowledge engineer (KE). A KE is an individual who is expert in structuring other's expert knowledge into rules and other formats that can be used by the expert system's inference engine in solving problems. This is a new field, now primarily staffed by computer

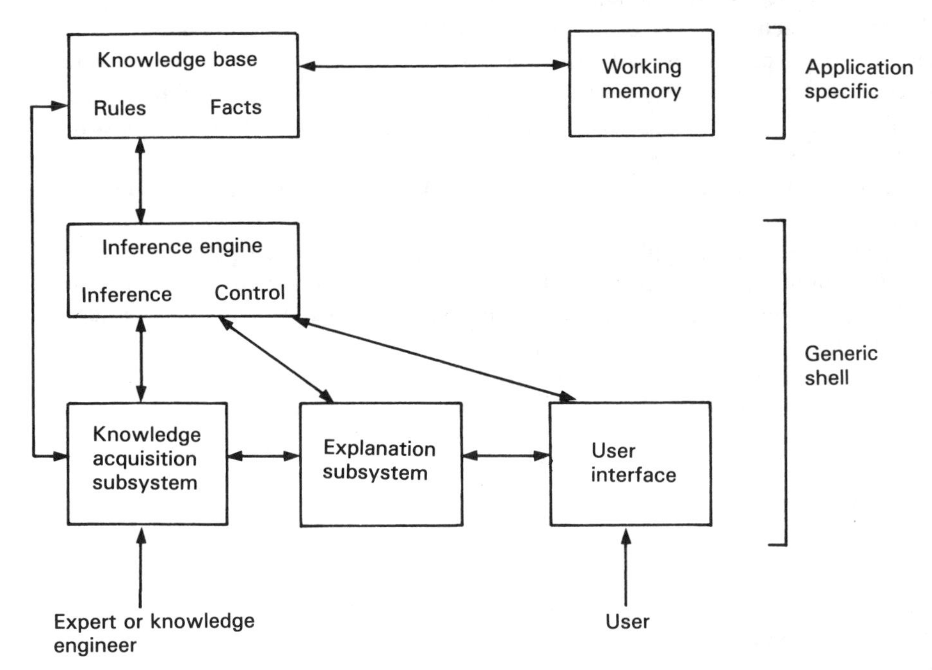

Figure 10.1 Architecture of a knowledge-based expert system. (Adapted from Paul Harmon and David King, *Expert Systems*, John Wiley, New York, © 1985, used by permission of John Wiley.)

science engineers. The KE works with the expert to structure his or her way of solving problems in a manner that the expert system will try to emulate. This is usually the longest segment of developing an expert system, and by no means is there a guarantee of success. The developers of the General Motors Charlie Program say they took over a year of close association with the experts to create a suitable working memory.

Developing a working memory requires that the often instinctive knowing what to do be broken down into

Problem definition
Symptom
Root cause
Elimination of root cause (solution)

This is analogous to the scientific method explained in chapter 6, which, you will recall, can take many iterations until the problem is resolved to the necessary level of satisfaction. In many cases there may never be a totally satisfactory explanation as to why an "expert" does what he or she does. For example, in tightening a tuning pin in a piano pin block, the skilled artisan may say he or she tightens until he or she feels the wood grip it. How do you define that to a computer program? Obviously, the artisan who has been doing the job for years knows what is meant and can repeat the action successfully over and over again to near 100% perfection. But, what of the repair person in the field called to adjust a piano and having to replace a pin, which is a relatively rare activity? How does the repair person know when the wood grips the pin? What can the expert system tell him or her to do? This is typical of the dilemma faced by KEs when dealing with experts and is particularly rampant when working with non-technical experts. If the ingenuity of the KE cannot find an alternative way of defining the problem, symptom, root cause, and solution, then an expert system cannot be developed. This is why creating the working memory can take such a long time and is never a sure bet to be successful.

The degree of accomplishment in describing the important processes an expert goes through in solving problems will determine how successful an expert system will be. This is a prime reason why these systems remain limited in scope at this time. The effort to debrief one expert can be enormous. Compare this with debriefing literally hundreds of experts involved with running the major components of a CIM system and then one begins to understand the enormity of the problem and why it will not happen soon.

Knowledge Base

This is where the KE structures the considerable amount of information obtained in the working memory into a format that the inference engine can work with. The important thing here is that the KE really understands what the expert has told him or her. Here we have an iterative process. The KE has to transcribe what the expert said into "computerese." In other words, into logic statements that will make sense to the computer. But, to do this with accuracy, he or she has to review each statement with the expert. From a practical viewpoint, to be successful, the KE has to become an expert in his or her own right in the field for which he or she is trying to develop the expert system. Vice versa, the expert has to become conversant with the style of computerese employed so he or she can check to make sure that the statement being employed in the knowledge base accurately reflects how he or she goes about making decisions. Now, if you think this is an easy mutual task, you have not had real world experience trying to merge dissimilar disciplines into a successful operation. Many people say this is analogous to a French person trying to teach French to an American while the American is trying to teach his or her counterpart English when neither speaks the other's language. It is probably not as bad as that, but it gives an idea of the degree of difficulty that can be encountered and why development time takes so long.

The KE resolves the expert's knowledge down to facts and the rules of how those facts are to be used. Example of facts follow:

1. The material of screws used for fuel injection systems is 314 stainless steel.
2. The material of screws used for motor flanges is c40 carbon steel.
3. The mean diameter of a sample of screws is 0.024 inch.
4. The mean diameter of a sample of screws is 0.46 inch.
5. Fuel injection screws are blue.
6. Motor mount screws are gray.
7. Motor flange screws are silver.

Rules related to these facts follow:

1. If the diameter of screws is between 0.020 inch and 0.030 inch, then it is used for fuel injection systems.
2. If the diameter of screws is between 0.40 inch and 0.50 inch then it is used for motor flanges.
3. If the color of a screw is blue, then the material is 314 stainless steel.
4. If the color of a screw is silver, then the material is c40 carbon steel.

This, of course, is a very simple example of facts and rules that can be contained in the knowledge base, and they are written in plain English for ease of explanation. A real knowledge base may contain a much more concise way of stating a fact and rule. For example,

Fact
1. Screw mean diameter = 0.024 inch
Rule
1. If diameter = 0.020 inch to 0.030 inch, then fuel injection system

This would be done to conserve database space and core memory space.

Now that we have rules and facts, we must decide what we want the program to do for us. The rules imply goals. Indeed, some practitioners say that besides facts and rules, there have to be goals. Others say this is not correct because to structure rigid goals defeats the purpose of expert systems and the whole exercise reverts to the rigid programming algorithms of the past. They say that, after all, expert systems are a subset of AI and, hence, the system must be capable of reaching an inferred solution, which means the goals have to be somewhat flexible. Suffice it to say this is confusing. So, just bear with me and shortly we will sort this out, and it should make some sense. For our example, let me state that the program ought to be able to evaluate any screw and, in an interactive fashion with the user, determine what kind of screw it is, where it is used, and what material it is made of.

Inference Engine

So far we have directed our attention to the application specific portion of the expert system. Now, let us consider what we can do with this meticulously structured knowledge base and what makes expert systems different than ordinary hard program structured systems.

The inference engine is the main component of the generic shell. This is the portion that can be purchased off the shelf. There are many commercially available expert systems shells that make selection difficult. One rule of thumb I think is useful is to use the shell with which the KE is familiar.

It is these shells that provide the semblance of AI through their capability of making an inferred solution to a given set of facts and rules. This is done through a series of chaining events, either forward or backward. This is the so-called inference programming indicated in Figure 10.1. Inference programming, along with the necessary control programming, make up the heart of the inference engine. However, it is backward and forward chaining capability that makes the shell what it is. Chaining is a means of using "and/or" logic to traverse a path to a logical solution. Figure 10.2 demonstrates this logic.

The logic is very direct. In forward chaining, we start with observed or known data and proceed through a series of logical steps to reach a most probable conclusion.

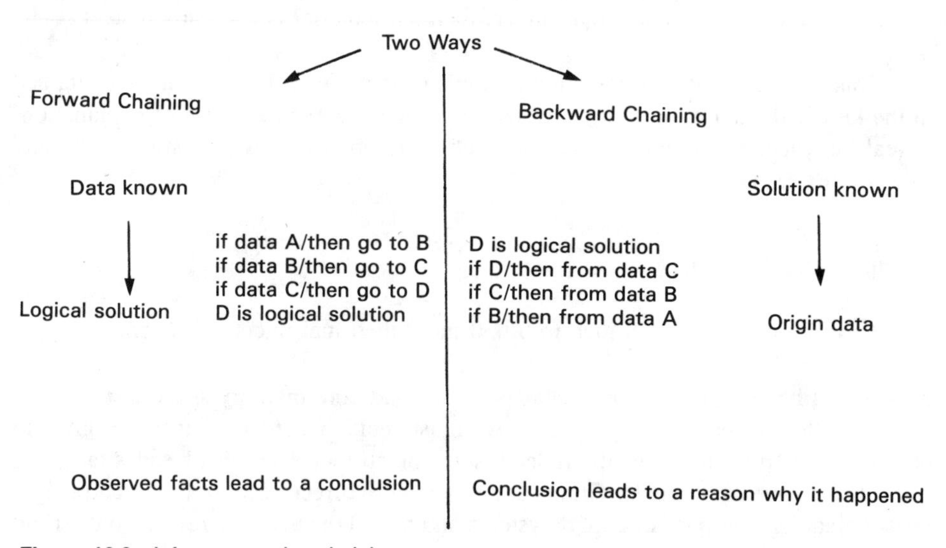

Figure 10.2 Inference engine chaining.

Many say this is the mechanization of the scientific method. In backward chaining, we play the detective game. A result is known, and we want to know why and/or how it came to be. Forward chaining has "if/then go to"–type statements, while backward chaining has the reverse "if/then from" statements. Both methods are valid variations of classical reasoning. Backward chaining is classical deductive reasoning, and forward chaining is classical inductive reasoning. Both offer a solution by relating facts to plausible chains of events. It is the need to infer that makes this type of programming different from ordinary computer programming.

Our computer heritage is primarily based on relating everything to numerical quantification (e.g., being able to do additions and subtractions). With numerical quantification, there is nothing to infer. The results are mathematically pure. There is no debate. MRP II is a perfect example of numerical quantification. Schedules are mathematical exercises and nothing more. An inference engine is not pure mathematics in a normal sense. It deals more in and/or logic than in mathematical logic, which means the answers may not be 100% correct. They are of a high probability of certainty, but not certainty. This is the area where we say expert systems are closer to normal human thought patterns (although humans are certainly capable of thinking in pure mathematical patterns too) and, hence, appear to be intelligent, which expert systems are not.

Figure 10.3 represents the simple logic path associated with forward chaining. Note that we start with a hard fact: a seed is planted. The logic takes us through a series of logical events that "probably" need to happen for a flower to bloom as a result of planting the seed. Note that in this chain of events there are plenty of areas for uncertainty to be considered. For example, based on the given facts, how do we know that enough hours of sunshine will occur this season for the plant to carry out enough photosynthesis to create enough food? Sufficient food is required so it will mature and reproduce by blooming. We do not know because we do not know enough about the genetic makeup of the plant and many other reasons. Therefore, all we know is that most probably the conditions will be sufficient for a significant proportion of seeds to evolve eventually into flowers. We know this through experience, and we know if the steps listed are missing, then the probability of flowering is nil. So, we have an inferred solution, and for most cases we are 99.9% assured of correctness. Figure 10.4 illustrates the same inferred solution but derived by deductive reasoning.

In both Figure 10.3 and Figure 10.4, if we were to program the sequence such

if a seed is planted	—data A
and	
if the seed is watered	—data B
and	
if the seed is fertilized	—data C
and	
if the sun shines enough hours	—data D
and	
if the temperature remains above freezing	—data E
then	
the flower will bloom	—logical solution

Figure 10.3 Forward chaining.

the flower is in bloom	—known solution
then	
there must have been a temperature above freezing	—data D
and	
there must have been adequate sunshine	—data C
and	
there must have been fertilizer applied	—data B
and	
there must have been adequate water	—data A
and	
a seed must have been planted	—origin data

Figure 10.4 Backward chaining.

that we had an expert system, we would program in a language that is based on and/or logic (recall Boolean algebra and the Venn diagrams) instead of the plus/minus logic of common programs. Instead of a precursor event leading 100% to a succeeding event, we would need some sort of logic that allows for a minority opinion no matter how low the probability of its occurrence. Therefore, we would have statements such as

If a seed is planted and watered or not watered then. . . .

This is an "if and/or then" instead of the more familiar "if/then" statement. This gives us other paths to consider in our search for the correct solution, which is probably more like the real world than a plus/minus-type of approach. The reason why we say expert systems need computer languages that consist of "if/then probable then" statements is because we want to consider the uncertainty in making our decisions.

Figure 10.5 illustrates the programming flow model of our flower forward-chaining example. Here we see a reason path and an unreason path. If we know the logic of success for planting a seed is a mature flowering plant (which has to be supported by the knowledge base of facts and rules), then the reason path is the one that takes us to that conclusion. But, there have to be countless other paths that could happen but not result in a flower.

In this figure, for simplification purposes, I have chosen to show every other path as a dead end. If the seed is not watered, what happens? Nothing. The plant dies, and we have no flower. So, for every "if," there can be many "ands" and "ors." The computer is instructed to follow the most likely scenario. It does this by having the knowledge base of rules and facts to refer to, and if it cannot determine a relationship based on given facts and rules, it interrogates the user to gain more facts and rules. In this process, the computer follows the "and" path(s) to the very end (e.g., running out of facts and rules and not gaining any additional information from the user). It then constructs a conclusion that is presented to the user. This conclusion will be as valid as the rules and facts supplied (garbage-in, garbage-out still applies).

With anything beyond the simplicity shown in Figure 10.5, for every "and/or" we usually place probabilities of correctness next to them so the program can calculate the degree of confidence we have in the solution presented. This is important because for

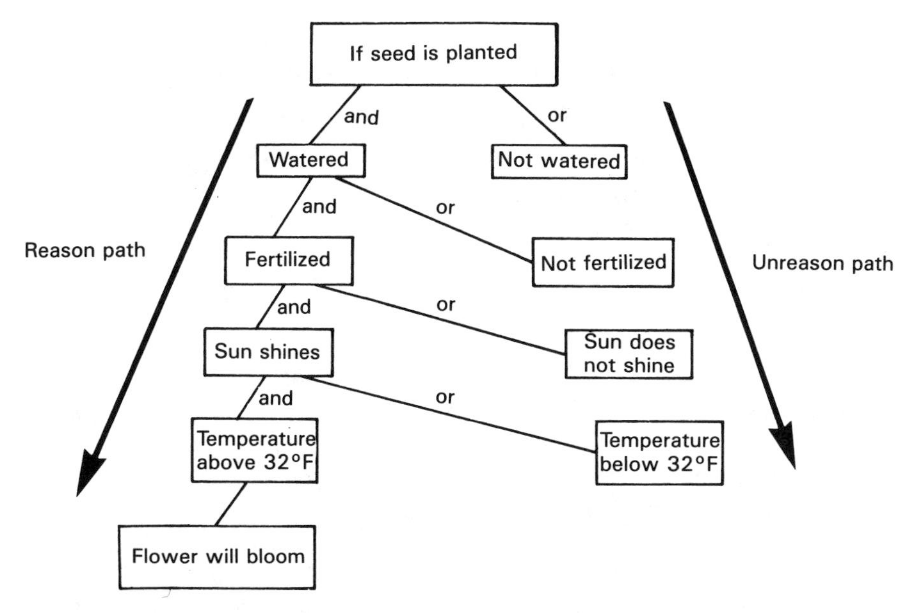

Figure 10.5 Simple and/or logic.

complex programs there is likely to be more than one possible solution, and it is necessary to distinguish which one is the most probable.

Now that we have investigated the basics of what an inference engine does and how it works, let us go back to the rules and facts we supplied to our knowledge base previously. In this example we have a set of seven facts listed and four rules listed. From these two sets, using chaining logic, we can infer some solutions. For example,

Fact
Fuel injection screws: 314 stainless steel
Rule
Screw is blue then material is 314 stainless steel
Inferred solution
314 stainless steel screws are blue and used for fuel injection

Another example is this one:

Fact
Screw mean diameter = 0.46 inch
Rule
Screw diameter between 0.40 inch and 0.50 inch used for motor flange
Inferred solution
0.46 inch screws are for motor flange

We can also continue and chain from this to get other information if we desire. We have other facts and rules about screws that will tell us more about 0.46 inch screws used for motor flanges.

Fact
Motor flange screws: c40 carbon steel
Rule
Screw is silver then material is c40 carbon steel
Inferred solution
c40 carbon steel screws are silver and used for motor flanges, and are 0.46 inch diameter

Here we have a case that shows the expert system "learned." It had already inferred from the previous example that the 0.46 inch diameter screws were used for motor flanges; hence, by chaining, it then decided that an "and" condition existed so the new inferred solution should include the previous portion "0.46 inch screws are for motor flange." Is this entirely correct? As far as we see with the data given, yes. But this does not presuppose that we could not in the future start out with, say, 0.42 inch diameter screws and going through the same set of rules come up with the conclusion that motor flange screws are 0.42 inch in diameter. This is entirely correct by facts and rules, but the program infers an exact solution, which is not so.

c40 carbon steel screws are silver and used for motor flanges, and are 0.46 inch diameter

We see this is a true statement but not an entirely true statement. Substitute 0.42 inch for 0.46 inch, and the modified statement is still true but not the whole truth. This is an anomaly from which expert systems suffer. The systems can hardly know the whole universe of truths for all conditions and are only as good as the facts and rules presented and what can be logically derived from those rules. There is no way the inference engine can grasp nuances to rules and facts that a person can. If this is kept in mind and it is understood that expert systems will always give literal inferred solutions (that sounds like an oxymoron but is not), then the systems have a use in limited applications decision making.

This shows that expert systems are capable of "learning." But we must be careful to understand what that means. This does not imply learning in a human sense. People learn facts and are able to use them in abstract ways. For example, I can say that forward chaining is analogous to the scientific method, and you, the reader, would understand that developing a hypothesis and conducting experiments to see if the facts support the hypothesis is vaguely similar in manner to defining "and/or" statements and testing them with given facts. I have also stated that using N/C machines is like having the best operators performing work on manual machines. Both of these examples make sense to us, but not to the computer operating an expert system because there is no direct relationship exhibited. Expert systems need direct cause and effect to "learn" and will not be able to extend that cause and effect unless it is given other related information. In other words, the computer cannot create "what if" scenarios. Humans only need experiences that might be tenfold to a hundredfold or more removed to connect items together in a cause and effect to learn. People can create "what if" scenarios ad infinitum with only the smallest linkage to previous learned information. We call this abstract thought. Computers have no capability of abstract thought.

The inference engine does possess learning capability in a computer sense. It has the ability through chaining to make interesting cause and effect relationships. This is not possible through hard programming because all the if/then statements have to be hard programmed in beforehand. Looking at Figure 10.6, we can see that the inference engine is capable of adding new facts and rules to the knowledge base to "learn."

Figure 10.6 shows that the user inputs gray screws. The software then searches the knowledge base for facts and rules about gray screws. It sees by relationship that gray screws are for motor mounts but there are no rules about gray screws. Here, the generic shell is preprogrammed to obtain further information by questioning the user.

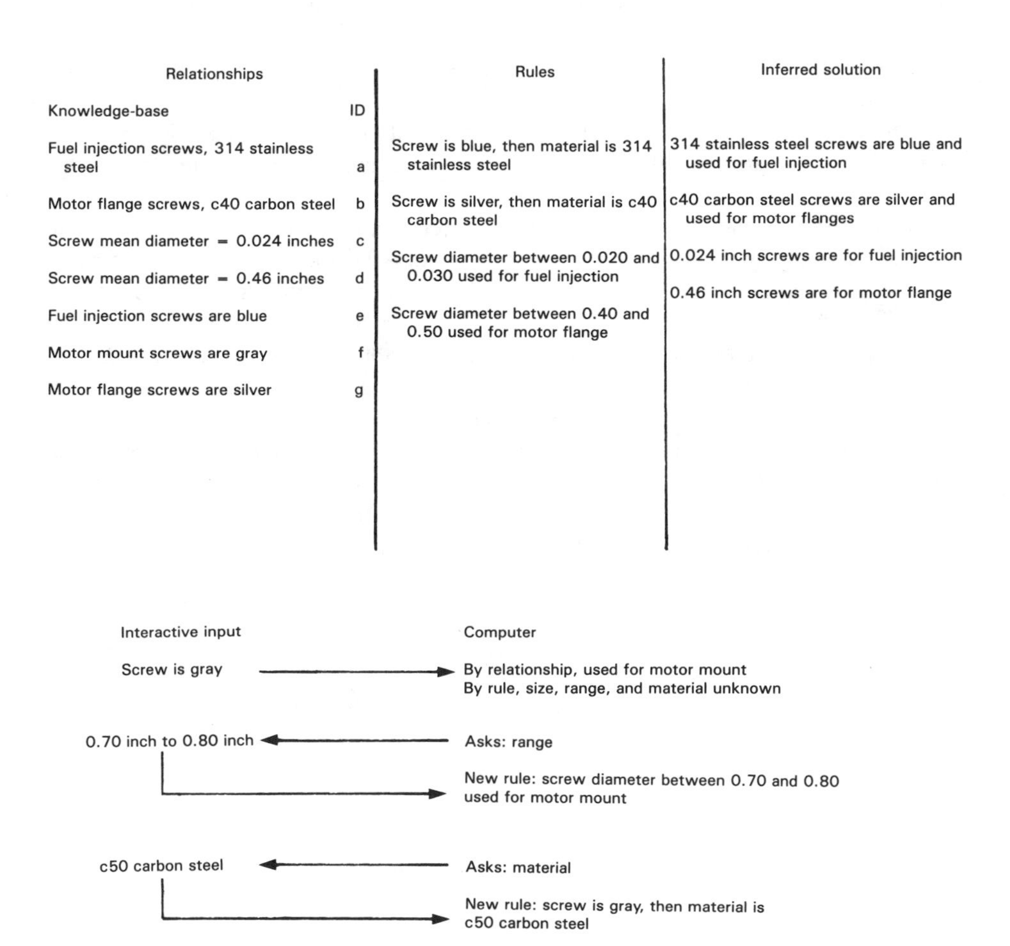

Figure 10.6 Inference engine builds new inferred solution based on experiences.

If it knows a fact, then it will ask for information pertaining to that fact. Vice versa, if it is given a rule, it will ask for information associated with that rule. In our example, it asks for information by the chaining logic to come up with meaningful relationships. In the example it is chaining through relationships to gain information on ranges and materials. It is doing this because its knowledge base has similar facts and rules. It has a fact relationship but no rule-type relationship, so it asks for range. It knows by experience that range and color are related. It gets a reply back that the range is 0.70 inch to 0.80 inch. Then, by chaining for previous associations, it connects this range with motor mounts to concoct a new rule as if the user put it in the knowledge base to start with:

screw diameter: between 0.70 inch and 0.80 inch used for motor mount

Next, it could ask for more information: material. The user would respond, "c50 carbon steel," and again the program would devise a new rule:

screw is gray then material is c50 carbon steel

The computer would ask the user if these additions to the knowledge base are acceptable. If they are, then they would become permanent additions to the knowledge base.

As implied throughout the discussion of inference engines, the software is different than traditional database programs. To use an analogy one more time, we can think of expert systems software as the relational type by comparison with the network type of traditional software, if we were analogizing with computer internal architectures. Recall that relational databases are very flexible and are capable of allowing ad hoc query of the database. Contrast this with networking databases that, while very efficient, are not easy to query with ad hoc requests. To do so requires writing of sub-routines and then debuging those routines. So, expert systems programs are analogous to relational databases while traditional programs are analogous to network databases. Expert system software is flexible, while traditional software is inflexible.

I do not think a detailed explanation of the software differences between expert systems and traditional programming is necessary in a text dedicated to managerial understanding of CIM. There are many texts and scientific papers available that go into excruciating details about the software theory for expert systems and its development. For those interested I suggest a library search of the American Society of Mechanical Engineers (ASME), the Institute of Electrical and Electronic Engineers (IEEE), and the Institute of Industrial Engineers (IIE) transaction journals for more specific information. However, as managers, we may need to know the prime generic differences so we can obtain the suitable resources in running CIM-based enterprises. In obtaining resources to implement expert systems, we will unfortunately find there are few qualified knowledge engineers who have project management experience as well as computer science expertise. It may be necessary to hire trainable KEs who understand expert systems and then educate them in the meaning of CIM and how to manage project implementation. The generic differences between traditional software and expert systems software may be a sufficient starter in finding the necessary personnel. I offer a generic difference, for starters, in Table 10.1.

What I have described, in layman's terminology, is how inference engines work. It is apparent that the inference engines have significant potential for improving applications of CIM. I will not spend much time describing the bottom three boxes of Figure 10.1, but I will go on to show how expert systems could be used to enhance some CIM modules. Suffice it to say that the last three boxes are the administrative modules necessary to effectively use the generic shell (sometimes called the empty shell) with an application-specific knowledge base.

Statistical quality control (SQC) appears to be a likely benefactor of AI-type programming, specifically expert systems. In SQC we have a generic portion that would apply to all users and a specific portion that would be particular to each individual user. Let us briefly review the characteristics of general and specific pertaining to SQC and then look at how expert systems logic could be applied.

All SQC programs are based on normal curve theory. This means that points of a universe of data should fall randomly and that two-thirds of those points ought to fall about the mean (a nominal 1 standard deviation spread). We also know that if the data are obeying normal curve theory, 99.7% of all points will be within the area described by the curve itself. This is defined as the area bounded by the 3 standard deviation spread. Therefore, if a process is in control, all data points ought to behave in accordance with the theory. If not, then the process is said to be out of control. The task simplifies to determining whether or not the normal curve theory is being observed on the basis of samples taken of the process. Regardless of the process being observed or the product or service being offered, this is what SQC generic routines are supposed to do. This is the case whether or not the details of making the in control or out of control decision is done with a computer.

After determining that a process is out of control, SQC programs in essence say, "You've got a problem, people, but don't look to me for help. I've done my portion to point out that a problem exists. Now you're on your own to figure out how to fix it."

Table 10.1 Generic Differences Between Expert Systems and Traditional Programs

Program	Languages	Structure	Primary Logic Philosophy	Control
Traditional	Basic Cobal Fortran Pascal etc.	Rigid Program steps executed in prearranged sequence	Mathematical (+/−/×/÷) if/then	Algorithm built into the specific programs
Expert system	Lisp Prolog etc.	Loose Program steps executed in an ad hoc manner depending on circumstance	and/or (chaining) if/then probable then	Inference engine based on rules and facts found in a knowledge base

Figuring out what to do is the specific portion of the SQC routine. Up to now there have not been sufficient computational disciplines to aid the operators, engineers, and managers in correcting the SQC discovered problems. They have been left to their own devices to determine causes from symptoms and then to design and implement a corrective action.

Both the generic and specific portion of SQC applications can be done more effectively with properly designed expert systems. The generic portion can be built into the SQC program itself with very little individual tuning for the targeted situations. If that were all we needed for a sufficient expert system for SQC, then we would hardly need a separate knowledge base. The specific portion, however, would require considerable expert information fed into the knowledge base, so a unique knowledge base allied with the SQC program is probably required. Starting first with the generic portion, what follows could be the basis for knowledge base facts and rules:

SQC generic facts

1. The mean for the process. . . .
2. The standard deviation for the process is. . . .
3. The upper control limit for the process is. . . .
4. The lower control limit for the process is. . . .

SQC generic rules

1. If the last seven data points are nonlinear, then process is random.
2. If process is random, then make no corrections.
3. If the last seven data points are linear, then process is nonrandom.
4. If process is nonrandom, then make corrections.

This is simple. It would probably take a KE only hours to properly format these thoughts into knowledge base facts and rules that an expert system shell would understand. The next part, coming up with the diagnosis of the problem and what to do about it, is the hard part. This is what will take time and require the process as described previously. Here, the interaction of the KE and the experts will follow the iterative process and many trial and error solutions before success is achieved. Unfortunately, that success will only be viable for the specific process. There probably will not be much transference to other SQC applications unless they are strikingly similar.

The generic part, similar to what was just described, is most likely commercially available today. I have seen trials of programs of which software vendors were doing marketing studies to see if the programs were commercially viable. Most likely the programs would be combined with a regression analysis package for testing data to see if the data were "mathematically" random or nonrandom and linear or nonlinear. The output would indeed be the progeny of an expert system, but, let us be honest, it would be of very little value by itself. To pronounce whether or not the process is or is not in control by means of an expert system technology is "super kill" and not worth the effort. However, if it is tied with recommended corrective action based on the current specific conditions, then we have something of value. This requires a specific knowledge base, which cannot be trivial. It needs to really be specific.

It is possible to concoct generic solutions to specific problems that would have reasonable chances of mitigating production losses, such as

If process is drifting out of control, shut down and look for cause.

But this is not a sufficient solution. This is non-use of experts and perhaps an easy way out resulting from not being able to discern what the expert meant. We would not be telling operations what to do to get the factory producing again. The need is for optimum solutions. Where should the operator look first, and why? What does "cause" mean? What are suitable suggestions that minimize downtime to the least time possible? Generalized specifics are not the solution. The need for a rigorous development of a specific knowledge base is required for expert systems to be useful for SQC applications.

What should this specific knowledge base contain? Unfortunately, we cannot say unless we know the specifics of the process. We can, though, state, with some degree of confidence, what the basic format should be, what categories of questions should be addressed. With this in mind, a format for an SQC specific knowledge base could be

SQC specific facts

1. Process correction x is. . . .
2. Process correction y is. . . .
3. Process correction z is. . . .

SQC specific rules

1. If nonrandom, then determine if drifting toward upper or lower control limits (note: we are assuming that the regression analysis program will tell which way the process is drifting if queried).
2. If trending toward upper control limit, then process correction is x.
3. If trending toward lower control limit, then process correction is y.
4. If two-thirds of last seven random points are not centered about the mean, then process correction is z.

Note that the specifics are specific! The process corrections are delineated as facts, and the rules are very specific as to when to use each remedy. This is the format that an SQC expert system should take if it is to be more valuable than one that just points out that a problem exists. Expert systems can be of sufficient value for SQC applications, and SQC applications should be looked into. But, keep in mind that SQC applications will take significant resources to develop into practical programs.

SQC should fall into the range of capability of successful expert systems. Expert systems are said to be deployable for problems that have relatively limited knowledge scope and those that would take an expert a few minutes to a few hours to resolve. SQC falls into these domains. Process control should be limited in scope. We do this on purpose. We install a specific process to do limited things to raw materials and assemblies in making products. We choose to narrow our choices of what is to be done so we can gain mastery and control. We want to maximize our capabilities and thus limit the choices of what we are assigning the process line to do. This way the operators of the line become expert in their tasks, and we optimize the output. If we limit the types of activities the process line does, then we also limit the expertise necessary to manage and troubleshoot the process. This falls precisely into the domain of success with expert systems, so it is quite correct to suppose that process control by means of SQC is a good choice for development activities.

Now, what of computer-aided process planning (CAPP)? Is it a good choice for development of an expert system? At first glance, probably not. But, if we use techniques such as computer solid modeling and group technology (GT) as an enhancer, perhaps a simplifier, then maybe we can do something useful.

The goal of CAPP is to use the computer to optimally select the best method and routing to make the desired part. The state of the art is variant planning. Here we use the software with GT coding to find similar to or exact match plannings to simplify the task of sequencing parts through the factory. If similar or exact matches are found, we need do very little additional work to get planning input to the MRP II scheduling algorithms. With exact matches, there is literally no additional work. The variant method only requires that total planning work be done if the part to be made is totally new to the factory. The generative mode goes even further by supposing that all the information necessary to make the part is resident in the common database (perhaps not in the planning database, but bits and parts are in the design, facilities, N/C, scheduling, maintenance, and other databases) and expert systems can be employed to pull all this information together to create optimal planning. Generative process planning needs some form of AI (e.g., an expert system) to be successful. No true generative planning system exists today, but what has to be done and a potential path to success is generally known and being worked on at many research institutions. Let us look at this process and see how expert systems may be expanded in the future.

As we have seen with previous examples, the expert system will need to have a knowledge base constructed for it. For generative process planning, it will have to be built about the following criteria:

1. A complex decision logic about

(a) The design of the product.

(b) The method sequence of building it. This means what operations can or cannot be done before or after one another. Example: welding has to be done before heat treatment, which has to be done before machining; otherwise, dimensional stability cannot be maintained.

2. Manufacturing knowledge of all processes the factory is capable of doing. This includes cycle times and facilities necessary to do the work.

3. The scheduling priorities for loading each work station.

4. The quantities of each type of facility and how each is currently loaded. This means the expert system (or a more advanced AI program) has to have dynamic information as to how the shop is loaded. Otherwise, the system is liable to erroneously plan for optimal work station sequences that cannot be used. This, as you can imagine, is not a trivial matter.

The expert system will have to be fed up-to-date information on all the procedures used in the factory. If this information is not kept current, then any planning the system develops may not be applicable to present situations faced by the plant's management. This is a different tactic than current expert systems. Currently implemented systems rely on the expert knowledge to fix problems that imply years of experience to find the best corrective action approach, for a generative planning system history is not all that important. What one did to schedule and dispatch to manual lathes based on qualified operator availability is not applicable if all those lathes were replaced with N/C machining centers. This implies that the half-life of a generative planning system expertise is going to be relatively short and must be updated constantly. One way of doing this will be to constantly draw information from the common databases pertaining to design upgrades, machine processes, method changes, machine tool processing sequences changes, and all other parameters that lead to work station loading decisions.

We can see that generative planning has to be dynamic to be useful. To date, expert systems take too long to build, which means generative planning systems are a long way from practical reality. We have discussed the problems with the KEs and the experts' being able to relate to one another to create adequate knowledge bases. We just will not be able to wait a year to develop a usable program. This means the ability to build knowledge bases has to become more mundane and clerical than innovative engineering. I relate the knowledge of how to build knowledge bases to that of N/C programming in the late 1960s and early 1970s. At that time it took the equivalent of a computer science engineer to define the program steps necessary to machine shafts on N/C lathes. Thankfully, we progressed to the family of parts programming within a few years, and now we are at the point of operators and secretaries doing all the input necessary to run N/C machines. It took considerable work on the part of engineers to achieve this stage. But, without it, N/C today would be a mere curiosity rather than the backbone of industry it is. The same will have to happen with expert systems knowledge base building to make expert systems viable tools for industry.

The promise of CAPP in a generative planning mode has many obstacles to overcome. A major one is the need for very fast access to large numbers of different sets of data and ways to combine those data sets in many different ways very quickly. One way of achieving the sorting capability would be to apply GT. We can certainly use the coding capability to describe virtually any useful grouping and to help in searching the various databases for the necessary information. We will probably be working with several different expert systems at the same time, so the GT classification and coding capability would be a great assistance. GT could be a means of accessing knowledge bases to obtain the information necessary to achieve a particular generative planning. Without GT, we probably would not be capable of focusing the search on the necessary elements of the individual or groups of expert systems. GT code is already the grand integrator of information for CIM. Applying it toward finding, filing, and indexing knowledge base facts and rules related to planning algorithms would be a very natural extension of the code.

Much work has been done in defining geometric shapes by computers. This intensive look at solid modeling has employed some sophisticated mathematics to link these shapes together to form virtually any structure we desire. This has led to modeling very accurate renditions of manufacturing components for structural analysis. By being

Table 10.2 Generative Process Planning Relationship with GT and AI

Needs	Achieved by
Break part into shape primitives	GT code
Create machinery/process primitives	GT code/AI
Put primitives in sequence order	AI
Initiate plannings in sequence request	GT code

able to model solid shapes rather than line drawings (said to be hollow drawings), it is possible to do precise predictions as to how these parts will react to various stimuli when actually built. If the computer were to know the boundary conditions of the shape and were to know it to be a solid, then it will calculate much different stress, deflection, and strain values than if it were to think it is a hollow figure (which all drawings really are). The complexities of defining a hollow drawing to represent a solid are immense and difficult. Therefore, the work with solids has been so important in simplifying the procedure, to say nothing of the improved accuracy of the results. These shapes are called primitives, and we can catalog them in the common databases and use them when we need them. These too can be found, filed, and indexed by GT.

Solid primitives have very distinct advantages for manufacturing, too. If we can apply GT, and then associate them with the finite catalog of what our particular factory makes, we can then create planning for building these primitives. Performing planning for simple solid shapes is much easier than performing plannings for completed drawings of a finished part. The primitive, let us say a cylinder, would only have a few steps because there are only a few basic methods for making a cylinder regardless of the chosen material. We would then code the primitives for their manufacturing information and for recall every time there was a need to make that primitive in relationship to a need for a complete part. This is much like the database used for variant planning. Building a catalog of forging primitives, machining primitives, and welding primitives, again simplifies the amount of data needed to be stored. We can devise rules and facts on how and where these primitives are to be used and we can end up with a scheme for a practical expert system for planning. This could be a way that generative planning would be introduced to industry.

We see that CAPP generative planning systems are more complex than the variant model. Generative planning will need AI along the lines of advanced expert systems to make them viable and also an intensive tie with GT. Table 10.2 shows these relationships and may be the order of development necessary to create practical generative planning.

I believe we have now discussed the major aspects of AI as it relates to CIM. AI, especially expert systems, has enormous potential for revolutionizing how we manage business functions. Whether the potential is fulfilled depends on the answers to the researcher's questions concerning what is intelligence and if a machine, a computer, can ever be built with intelligence akin to human beings. These are very thought-provoking questions, ones that I do not remotely contend I have the answers to. However, in our narrow domain of manufacturing business management we can see some very useful developments that could give us significant improvements in using our resources and becoming better managers.

Chapter 11

Networking Communications for CIM

CIM is "communication excellence." Therefore, it is logical to conclude that a company operating under a CIM philosophy will be employing superior communication techniques to link its various activities. We have explored numerous facets of CIM, and we always come back to the need for an integrated approach, one in which access to a common database is mandatory. In this chapter we will explore the integrated approach from the viewpoint of communications strategy. First, we will discuss the integrated approach for optimization and then discern what is important for communications success. We will then discuss various communications scenarios that will support the needs of the integrated approach for business success.

We have looked in some detail at the three components of the CAD/CAM triad: machine/process control, design and planning control, and production and measurement control. We have seen how each component emphasizes one aspect of CIM but requires input from the others for optimum success. This tells us that optimization of one component at the expense of the others leads to suboptimized organizations. Business abhors suboptimization. All examples and discussion so far have shown that the power of CIM is its ability to facilitate business's capabilities for optimum results. In fact, I have shown case after case of non-CIM solutions leading to less than desirable results. We also know that CIM is the integration and automation of the seven steps of the Manufacturing System. All this points to integration of efforts by an organization as the most beneficial way of solving any problems. This is teamwork, with a computer strategy designed to take advantage of the teamwork concept.

CIM is a business philosophy, not just a manufacturing philosophy. Integration is not limited to manufacturing but by necessity must be carried throughout the entire business entity. If teamwork is to exist, then this is as basic as the sun rising in the east every morning. Let us further define integration in a business sense. Integration means all resources of a company are brought to bear on the problem at hand. It transcends the needs of any one function for the benefit of the whole.

The integrated approach is the only rational approach. Even if CIM did not exist, this would still be true. Here is an allegorical example. We have a major league baseball team striving to win the pennant. It enters the free agent market to buy star players. The owner of the team, in order to get the stars to sign with his team, individually impresses upon them the possibility of becoming the league's most valuable player because he will be playing with so many other outstanding players. The owner does not mention winning the championship even though that is what the owner wants. The owner stresses individual performances and the lure of individual recognition. Imagine a team of star baseball players all entirely consumed with desire to become the league's most valuable player (MVP). To become the MVP, a player needs superior batting average, runs scored, runs batted in, and very few errors committed on defense. Note that nothing is said about winning games or successfully executing a sacrifice bunt to move a runner into scoring position so another player may drive in the run. That task, sacrificing, in the mind of the MVP candidate is left to players of lesser talent. If no one is willing to sacrifice for the good of the team, many opportunities for winning games may be lost. If too many are lost, the team does not win the championship. An odd thing about not winning championships is that very rarely is the league's MVP selected from non-champion teams, no matter how good a player's statistics are. With everyone striving to be the MVP and no one worrying about the overall goal of being a championship team, that team rarely comes in first. This is suboptimization. If you are a ballplayer it means not sharing in the prize to be divided among the winners. In business it means not being successful, and perhaps going out of business, if coming in second or third happens too often.

Now, an integrated approach does not mean automatic success, but it does mean your team has a much higher probability of success than a competitor who has a team full of self-serving MVP candidates. The owner in our allegory struck out. The owner wanted a winning team and thought if the best talent for every position was obtained, then this was a sure thing. However, the owner overlooked the most important element: superior teamwork. Since this was not stressed, the team did not win. The team watched the World Series on television instead of being there as their individual talent, if properly integrated, would have ensured.

In business, we too often strive to be the MVP instead of the team player. We build superior marketing, engineering, and manufacturing organizations and forget about integrating them. In the firms not operating with a true teamwork approach, we compound the error by allowing these functions to set their own goals virtually oblivious of one another. We have our MVP candidate-dominated teams that achieve suboptimal results. This is not satisfactory. This is why companies sometimes find they cannot compete with international companies dominated by teamwork philosophy. The integrated approach, even a non-computer-assisted one, is the only rational approach. Add in computers to assist in effectively integrating efforts and we have the superior approach offered by CIM.

If our MVP caliber manufacturing and engineering organizations are busy doing their duties independent of each other, and probably doing them in a superior fashion, then the net gain for the company is still probably nil. I can easily imagine manufacturing striving mightily to reduce production costs by improving work station efficiency and by minimizing quality losses through SQC, and hosts of other good things. They calculate all the fruits of their efforts, get an impressive sum, all to be negated because engineering changes the design, which now calls for more work in process steps, tighter tolerances, and more expensive materials. Working independently, each area may have achieved its internal goals. If individual subfunction goals were integrated through a properly conceived CIM communications network, the chances are the subfunctions would not have been working at cross purposes. We would not be seeing an all too common scenario: The right hand not knowing what the left hand is doing (e.g., an abject and total communications failure). CIM communications networks in common databases are designed to prevent this sort of thing from happening. If it were not so sad, it would be humorous: one subfunction (manufacturing) reduces costs while the other (engineering) adds costs, and both are stiffly virtuous in proclaiming they had the best interests of the company foremost in their minds.

What we see here is the traditional independent tactic for organizations. The leaders of engineering, manufacturing, and marketing think of themselves as independent entities loosely associated within the overall company. Then tend to want to operate independently because it is easier to do so. This is because in non-CIM environments, communication is difficult. It does not happen automatically.

Because non-CIM environments do not typically have good communication networks, the tradition of each function is to operate independently. They tend to optimize areas they have control over. The more shared responsibility dictated by the general manager, the less likely the functions will put significant resources into them. The reason for this is obvious. When dealing across functions, the communication networks are not thoroughly established. Hence, if communication is not sufficient, then there is a lesser chance that the entire activity will be successful, and no manager wants to be associated with failure. This results in the rational integrated approach being abandoned in favor of the irrational independent approach. To be rational, the managers have to have good communication systems. If this proves difficult to accomplish, what logic dictates ought to be done is thrown away for an easier but less optimal solution. CIM communications systems dominated by common databases make it easier for the managers to select the rational approach. With good communication, there is less risk, and it is akin to the function manager operating within his or her own sphere of control, so he or she is willing to dedicate resources to solving the problem. This is also possible because the common database, while common to all, is controlled for input and output of specific data by the prime users. Hence, the function managers can maintain control while working effectively and easily with other function managers. In our example this means we would not see manufacturing striving to reduce cost on a soon to be obsolete design. Instead, they would have been working to reduce costs on the design engineering was about to introduce.

In the traditional mode, functions are forced to operate independently. This means their goals are truncated, which in turn means that they cannot be expected to achieve as great a return on the investment in their activities as an integrated team of functions can. Traditional mode functions do not react quickly to changing environments that are not of their own making because they suffer from communication lag. CIM philosophy requires excellence in communication. If a CIM system is in place in a company, the communication networks allow excellent dispersal and receipt of information and make the integrated rational approach a practical reality. Let us look at another example. If we have a CIM communication network linking databases, the following scenario about the effect of a tolerance change is entirely feasible:

- Engineering changes design tolerance. This is communicated to the common database by means of the communication network.
- Manufacturing receives the communication by means of the network and then enters the common database to evaluate the tolerance change impact on processes needed to make the product.
- Manufacturing as a result needs to change cycle time. This will increase costs. Information is entered into the common database by means of the communication network.
- Finance receives communications by means of the network, keyed by the product cycle time change.
- Finance calculates cost changes and then enters them into the common database by means of the network.
- Engineering reviews cost changes from the common database communication network to see the effect of the tolerance change. They determine if the change is worth the extra cost.
- Manufacturing reviews cost changes from the common database communication network. They determine what additional process refinements can be made to minimize the changes in cycle time and then inputs these modifications to the common database communication network.

This cycle is repeated over and over again, as many times as necessary until the change is thoroughly explored. This way the decision as to whether or not to change tolerances is made with real information and full participation. Chances are that the decision reached in this way is far superior to one arrived at unilaterally. The beauty of the integrated rational approach by means of CIM is that these cause-and-effect changes happen very fast and in a visible manner. Since changes are visible, they are immediately open to praise or ridicule. Everyone enjoys praise, not ridicule. So CIM tends to elicit better quality work from all functions because change agents cannot hide the results of their work. The avoidance of damnation is a powerful motivator of superior effort. How does this come about? It comes about through excellence in communication, by establishing a synergistic set of activities within all functions of the company. The results are profitability improvements greater than the sum of the parts.

The integrated rational approach dictates a direction for the designers of the supporting communication system (see Figure 11.1). This approach, while different in

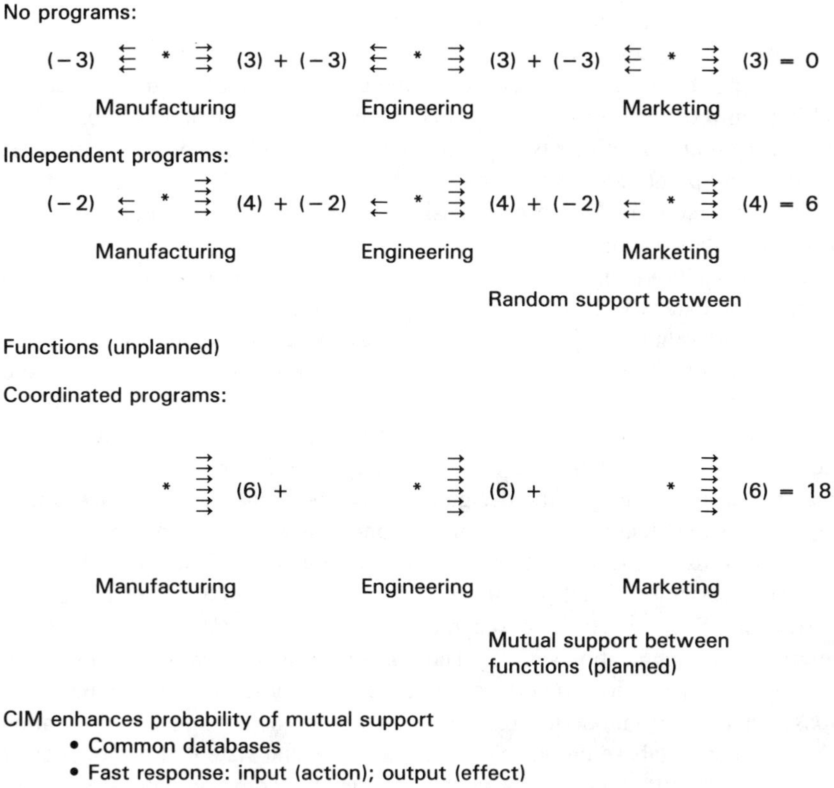

No programs:

(−3) ⇇ * ⇉ (3) + (−3) ⇇ * ⇉ (3) + (−3) ⇇ * ⇉ (3) = 0

Manufacturing Engineering Marketing

Independent programs:

(−2) ⇇ * ⇉ (4) + (−2) ⇇ * ⇉ (4) + (−2) ⇇ * ⇉ (4) = 6

Manufacturing Engineering Marketing

Random support between

Functions (unplanned)

Coordinated programs:

* ⇉ (6) + * ⇉ (6) + * ⇉ (6) = 18

Manufacturing Engineering Marketing

Mutual support between
functions (planned)

CIM enhances probability of mutual support

- Common databases
- Fast response: input (action); output (effect)

Figure 11.1 Benefits of integrated approach to productivity improvement.

specific application, will have generic similarities imposed by the very needs of integration and for providing the capability of all functions to have equivalent access to database-contained information. The integrated approach requires all users to be linked together, almost like a cozy rural telephone system party line.

If we look at Figure 4.2, we see the various steps of the Manufacturing System and the supporting hierarchical computer systems. We see that all the computers are linked together to form a common database. What we do not see is the companion communication system that is capable of accessing the entirety of the database regardless of the point of inquiry (e.g., work station on the factory floor or terminal in production control or executive terminal in the vice president of manufacturing's office). Figure 4.2 correctly implies that driving information for the process of running the factory is different depending on the hierarchical level of the computer-driven equipment or work station terminal. This means that not all information (think of it as CAD/CAM triad information versus human control information) is allowed to all terminals, only that information based on the function's needs at that location. This is required to protect the various functions from being inundated with information irrelevant to the specific task. This is not the case for a communication system for management control. This communication system, different than the functional requirements system of Figure 4.2, has to have equal access for managerial communication from any access device. It also has to be capable of ad hoc query to get the user's information in a format that is acceptable to him or her. So we see the first generic requirement for an integrated rational approach communication system. Access has to be available in an equal manner regardless of access point. This is analogous to not allowing any restricted telephone lines. All lines are free to make those expensive long distance calls, and we trust the users not to abuse the system.

The integrated rational approach communications system is more than a way to access the common database. If that is all it is, then it would be virtually redundant to Figure 4.2. Recall that in the example of the desire to change a tolerance of the design we saw how the various integrated functions went back and forth to the database and were at the same time capable of querying each other. We see that functions received communications by the network. This means specific functional users either tapped the common database to see what was new (like saying, "Good morning computer. Did anyone change the information in the design database last night? If so let me see it.") or the originator of the change by means of the communication system (linked to the common database) instructed the system to literally ring up other functions. The other functions, upon receiving a signal, would then query the common database and get the latest information. So, we have the second generic requirement for an integrated rational approach communication system. There has to be an easy way of transferring information between the various functions, not necessarily or totally through the common database. In fact, most modern communication systems do not transfer information through the common database but instead send messages and state that if the receiver wants to know the details, then he or she should enter the common database at their discretion.

To be effective, messages have to be transferred quickly and accurately. This is required for all users, both for the functionality of performing CIM-driven integrated work (Figure 4.2) and the supporting integrated rational approach communication system. So we have the third and fourth generic requirements for an integrated rational approach communication system. There has to be a quick response time, usually three seconds or less. This is the so-called three-second rule. Psychologists tell us that if we have to wait longer for a response too often, we tend to get cranky and ornery. Hence, delays would not be good for our morale, to say nothing of our productivity.

Transmission accuracy is usually taken for granted because data transmission is typically good for local systems. However, nothing we have discussed says that a CIM system has to be a local system. If it is not, then transmission accuracy could be very important. This is recognized by the fact that the long distance communications carriers are most willing and anxious to sell "clean" line services. So, accurate transmission is the valid fourth generic characteristic.

A fifth generic characteristic is the ability to converse with more than one user of the system at any time. Very often in voice communications it is desirable to converse in a group of more than two. This is demonstrated by the many conference calls placed daily by large numbers of businesses. Quite often it is more efficient to have group or parallel conversations than a series of serial conversations. The same is true with data communication between functions, as we would expect.

We can summarize the generic requirements for an integrated rational approach communications system to be

1. Access available in an equal manner regardless of the access point.
2. An easy method of transferring information between functional users.
3. A fast response time, usually three seconds or less.
4. Access to the common database at will with no intolerable delay.
5. Capability of easy multi-user simultaneous communication.

Looking at Figure 11.2, we see the difference between one-on-one communication, the independent communication system, and the communication system required by the integrated rational approach, the integrated communication system. We see that with the independent flow approach each user is in essence an "island of communication" set up to support the old fashioned "island of automation." This is the antithesis of CIM. It violates the premise of integrated activity being stronger than the sum of the independent parts activities. Note that there are no shared communication lines between users and the common database. Granted, each has access to the common database and each can talk to each other, but none can easily talk to the entire group or listen in to what other members of the group are saying to each other. This is an awkward system. In fact, if another user joins the group, then one-on-one lines to the other four users and to the common database would have to be run. With an independent system we in effect have to place a series of independent message transmissions if the message is pertinent to more than one function.

For a communication system to be acceptable for the CIM-required integrated rational approach, all five generic requirements have to be met. It is evident that the independent communication system does not achieve this goal. It fails, in order of certainty up to highly probable, on points 5, 2, and 3. We can make a case that the independent communication system is satisfactory for point 1, equal access, and point 4, access to the common database with no intolerable delay. There is no way point 5, capability of easy multi-user simultaneous communication, can be satisfied in a reasonable manner. To do so, each user would have to be a full-fledged communication relay system. This is hardly what we mean by easy. Point 2, an easy way of transmitting data between users, probably cannot be satisfied. While one or two users, perhaps three, can be, think of the difficulty of sending the message to thirty users, or more, on a one-on-one basis. Thirty users in a CIM-implemented factory would not be unusual. Just the time needed to repeat the transmission to each intended recipient would be prohibitive even in our electronic world. This says nothing of the transmission costs. Point 3, fast response time, probably cannot be met either. For one on one it would be possible to communicate within the three-second limitation; but, for large numbers of users responding to a query, it is not feasible. The independent communication systems do not meet the requirements of the integrated rational approach, and we have to look elsewhere for the solution.

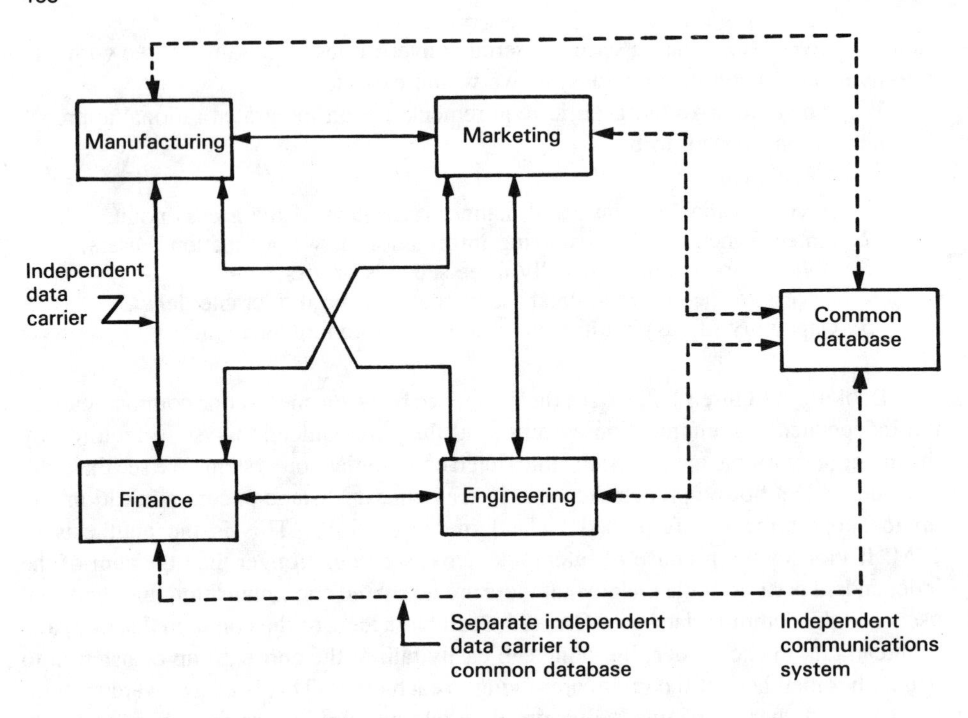

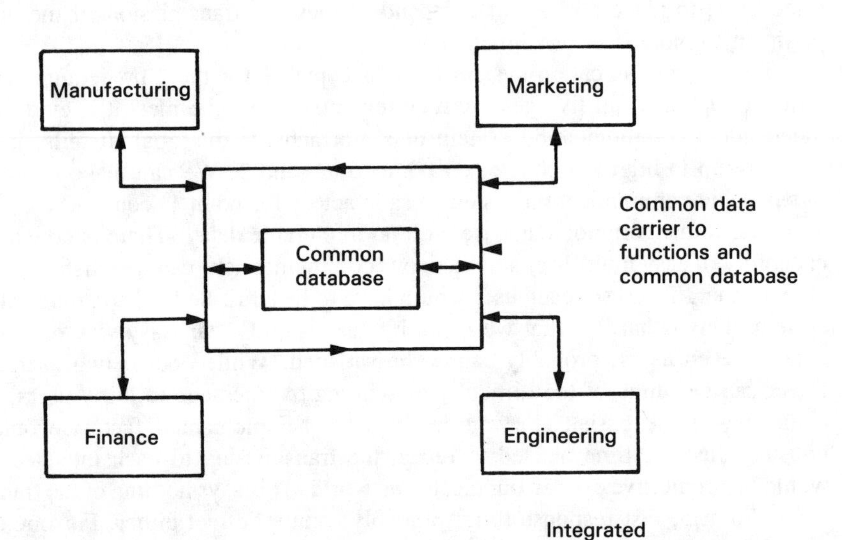

Figure 11.2 Difference between independent and integrated communications systems.

It did not take long to find a communication system to meet the requirements of the integrated rational approach. Our electronics engineers just had to look over at the other end of their profession's spectrum: power engineering. Here we see sophisticated and massive use of the theory of grid systems. Power generators, located far apart geographically, all contribute power to the grid simultaneously, thereby supplying the power needs of huge population blocks over hundreds, even thousands, of square miles. There are requirements for entering the power grid that all generators must meet. Primarily, alternating current phase angle has to be in synchronization and the voltage at which they are generating has to be the same. If this is so, the amperage or quantity of power is additive to the grid.

The same analogy can be used for a communication system. Establish a grid, with rules of entry, that in essence creates a common highway for all users to enter and exit at will. We call this a data bus. This is the way modern telephone systems are set up, so there is no reason to think it will not work for a database-linked communications system, and it does work. The rules are fairly commonsensical, even if they are surrounded by the mysteries of electronics. Like the power system, they have to be of the same voltage, or electromotive force, so they enter the highway at merge speed. In other words, our transmission has to be at the proper character transmission rate for the data bus to handle (e.g., so many bits of information per time period). This is commonly called baud rate. If it is slower or faster, then they get rear ended or bump into messages ahead of them, the result being an accident that ties up the system. So, the baud rate has to match. Instead of phase angle, as in a power grid, as a requirement, the communication systems typically require a frequency at a specific hertz (cycles per seconds) so every message is in synchronization and the system knows a message is in the circuit.

Like a superhighway or a power grid, a data bus is capable of carrying many messages, and they get to their destinations quickly with no loss of content.

We can see in Figure 11.2 that the integrated communication system links all users together simultaneously through a common communications network. Further, all five generic requirements for an integrated rational approach communication system are met. We have overcome the obstacle of how to comply with points 5, 2, and 3; and points 1 and 4 are still achievable. The common data bus ensures easy multi-user simultaneous communications (point 5), easy transfer of information between users (point 2), and a fast response time (point 3) regardless of the number of users.

What we have here is commonly called a network, many users hooked to a common information carrier. In fact, the integrated rational approach communication system generic requirements define any network requirements in the larger sense. This means it can be a voice communication network, such as a long distance or local telephone system, or even a television broadcast network, as well as a linked computer-driven factory system. The integrated rational approach dictated by CIM philosophy leads us to the definition of communications network: A linkage of devices that allows simultaneous receipt and transmission of information.

We will use the term network to define the linkage of terminals used for business functions to communicate in parallel to and with the common database. The network is our communication system necessary to manage the entire overall CIM system. Now, let us look at how these networks are constructed and what the future will require.

Now that we see why the concept of CIM allows and demands integrated approaches to productivity improvement, let us see how CIM has led to better and better communication devices, protocols, and procedures. We will focus on internal factory communication links (networks), but the discussion is generically valid for all communication links where there is a mix of cognizant computer and human interfaces.

In recent years a term has come into use to indicate that the network we are talking about is associated with the CIM strategy. That term is "Local Area Network" or "LAN." When this title is used, we automatically know it is a communications loop set up to meet the integrated needs of CIM. The commonly agreed-to definition of LAN follows: A communication system designed to transmit data from one point to two or more other points simultaneously within a relatively small area.

Note that the definition does not require that the points be computer terminals, databases, or any other electronic devices associated with operation of a CIM system, but neither does it preclude it. Further, the fact that we must transmit data to two or more points simultaneously denotes that the solution to the requirement will most likely be solved only by a common database-type of system similar to that shown as the integrated diagram of Figure 11.2. The desire for LANs comes from the desire to have a CIM solution to running a business. More succinctly, it comes about through the CIM need for distributed data entry and processing. We realize that for CIM to work, we cannot be bogged down by bureaucratic central data entry and processing facilities. Therefore, LANs answer the need to move those activities out to the user's location without losing any of the capability of high-power central processing sites. A corollary reason for the development of LANs is the increased emphasis on office productivity improvement to keep up with trends in factory automation. Office productivity improvement stresses the need to streamline the clerical function; in other words, progress toward the definition of CIM, which is excellence in communication. LANs provide for faster, more efficient, more dependable means of communication, user to user, office to office, machine to machine, and computer to computer.

How complex can a LAN be? The answer is very complex. They are limited to local areas, but not by complexity. The limit to local area is not a significant constraint for a business. Many surveys over a number of years indicate that 80% to 90% of all business communications occur between users and devices within one mile of each other. So, LANs become complex owing to the nature of the task of encompassing all of that activity. LAN communication devices connect points in the same room, same building, and the adjacent groups of buildings usually all within a mile of one another. However, it is not unusual to have a single or a few longer distance lines of up to five miles away. I am convinced that as communication technologies improve, the distance questions will disappear, and we will find LANs merging identities with normal telecommunications systems.

LANs support communications between any of a wide variety of devices. This may be thought of as another distinguishing factor between LANs and telecommunications systems. LANs must effectively link devices users had before the advent of the LAN as well as new devices. LANs must have the ability to add onto the system without degradation of service. This is another reason why the integrated diagram of Figure 11.2 and not the independent one is the only suitable model for LAN development.

Types of devices LANs have to link together include, but are not limited to, the following ones:

- Computers
- Data entry terminals
- Machine tool and process control units (MCUs)
- Word processors
- Electronic copiers such as printers and FAX machines
- Engineering work stations and drafting stations
- Telephone switchboards of all types
- Ordinary telephones

The development of LANs is relatively recent history, as we would expect. As CIM theory became more and more mature, the corresponding communications schema grew with it. LAN as we know it today dates to the late 1970s. I will briefly discuss two early LANs so the flavor of the technology that had to be developed becomes apparent. We owe much to these pioneers because they tackled the hard problems and demonstrated that LANs are not only practical but they offer significant competitive advantages to those who employ them.

Ethernet. This early version of LAN was developed circa 1980 by a consortium of companies including Xerox, Digital Equipment Corporation, and Intel. The network architecture provides for a single channel for digital transmission. The channel has a capability of transmitting one to fifty megabits of information per second to the hooked up communications points. There is no switching logic within the network. This means the users at their ends, independent of the network, need to have screening devices so only information of use to them is received.

Users enter the network in a manner similar to an automobile entering the freeway. The driver waits for the proper slack in traffic that allows him or her to accelerate to the highway median velocity. In the case of Ethernet, this means entering the coaxial cable with the message using the "Listen Before Talk" and "Listen While Talk" protocols. The Ethernet devices that transmit and receive have built into them electronic circuits that prevent transmission if another message is already on the coaxial cable. It will wait the micro- to milli-seconds until the message ahead of its message is cleared before transmitting. Also, while transmitting, it will listen for other devices trying to get on the line. Depending on the priority of the next message, it will either continue to transmit or wait for the higher priority message to go by, similar to a slow freight train pulling off to a siding to wait for a faster passenger train to pass. Because of the single channel, the limit to effective numbers of users is moderate. It is possible to degrade the service (speed) with this system if the circuit has many users and they all need to be transmitting at the same time.

Wang Net. This LAN was developed by Wang Laboratories about 1983. It differs from Ethernet in that it is an analog transmission system that takes advantage of the huge bandwidth available with cable television. The older Ethernet system used digital transmission. The difference between digital and analog is significant in how much information can be transmitted within a specific period. Digital transmission has numbers and corresponding letters expressed directly as characters of a binary-based code. Here, all letters and numbers have to be converted to a series of zeros and ones, where zero equals no voltage and one equals unit voltage. Once this is done, the on and off transmitter sends these pulses and the receiver then decodes them at the other end. The analog system expresses numbers and letters as directly measurable quantities of voltage. In some cases it interprets twists of a dial or changes of resistance as the variable and encodes that as voltage levels for transmissions. Analog is said to be more discerning because of the infinitely more variation it has as compared to a simple off/on transmission device. This way it can support simultaneous voice, data, and video transmissions, but it is more complex. While Wang Net has communications protocols to prevent miscommunication and garbling of data, it is less restrictive than the older Ethernet. Wang Net, running on the much wider bandwidth and using analog transmission versus digital transmission, has a much larger capacity. This means many more users can be hooked up and conversing at any one time.

Wang Net also covers a larger area than Ethernet. Wang Net, taking advantage of cable television long lines capabilities, can effectively stretch to cover areas up to twenty square miles. Ethernet is definitely a one-building communications network. With its single channel it would be foolish to extend it too far. Wang Net is an evolutionary improvement over Ethernet. It takes advantage of experiences gained by its predecessor and builds upon it. Today we still see both of these types of systems in operation. But, as expected, they have both undergone modifications and improvements. The debate as to which is better, analog or digital, goes on, and each is used as a selling endorsement. Those who favor simplicity as a goal tend to favor Ethernet-type systems. Those who need the wider capabilities offered by analog systems opt for the Wang Net–type systems. Of interest, where much scientific data are being transmitted, the analog system is preferred. This allows considerably more complex data to be obtained and transmitted to many interested recipients, usually for scientific analysis. Digital systems are preferred where voice communications predominate.

LANs were first introduced to satisfy a limited need, the need to have a communications system for a finite CIM system. However, it becomes apparent that CIM systems, especially those developed for multi-site, multi-division companies, need the ability to occasionally communicate to other CIM systems. This has led to a family of devices called gateway processors. Gateway processors are buffers between the LAN and the common carrier telecommunications systems, usually the telephone companies.

We can think of gateway processors as the electronic equivalent of the common step-down transformers we see on power line poles. The local power company transmits power at voltage and amperage to most efficiently prevent line losses. This voltage and amperage is too high to be safe for home or factory use, so it has to be stepped down. A similar thing happens in the LAN, only here we are concerned with line protocols. We have seen with the two basic types of LANs that they are significantly different in their approach; therefore, it comes as no surprise they do not match the protocols used by the telephone utilities. If the utility is going to be used to send information from one LAN to another, there is a need to make the information compatible with the carrier. A gateway processor does this. It is an electronic matching system dominated by a small computer that can translate LAN protocol into utility protocol and vice versa.

These are complex devices that have many switching devices directed by very specific software. The unfortunate fact is that they tend to be specialty design products, in that there has to be a direct match between the two networks with no loss of meaning in the translations.

The major problem is that there are still no recognized standards to be employed in the design of LANs to enable gateway processor designs to become standardized. This has led to the latest development in the maturing of CIM: the concept of communications standards throughout all the devices industries. All the devices a LAN is supposed to communicate with, and all the gateway processors, tend to be unique designs. Each of these designs is optimized to solve the specific problem at hand, regardless of future communications needs. Therefore, we really have a "Tower of Babel" situation. This has deterred full implementation of CIM, especially for large companies. We find that MCUs and terminals of different manufacturers cannot communicate with each other within proposed LANs. This dilemma makes CIM installation difficult. To overcome this, gateway processors have even had to be employed internally, which increases the cost of implementing CIM modules significantly. Further, it may even defer doing so because of difficulty in economic justification. It is understandable that gateway processors are needed between the common carrier utilities and the LANs because of the unique requirements of the LANs versus the utilities. But, there is no good reason why MCUs cannot communicate with terminals and computers made by other manufacturers. Recent attempts by technical societies and major industrial firms seem to be on the right track toward solving this problem.

The future of networking for CIM communications seems to lie with standardization of components making up LANs and devices needed to control factory processes. As we've seen, the "Tower of Babel" problem must be solved if CIM is truly to be universally applied and successful. We must all learn to speak the same language and use the same formats. Just as English is the standard language of civil aviation, we need a similar discipline for CIM. There may be such a discipline emerging now, and it is worth discussing.

MAP/TOP has gained much attention as a communications protocol standardization for CIM. If it is adopted, it will define the standards against which all LANs and communications devices used within CIM will be designed to. MAP/TOP means Manufacturing Automation Protocol/Technical and Office Protocol. It is proposed by General Motors and Boeing Aerospace, and has been endorsed by the major computer hardware producers such as IBM, AT&T, and Digital Equipment. The Society of Manufacturing Engineers and the Institute of Electrical and Electronic Engineers have taken on the responsibility to be a liaison with the technical community to develop, publish, and continuously update the standards. Also, considerable progress has been made toward obtaining acceptance by the International Standards Organization (ISO), the consensus standards organization subscribed to by the world's industrialized nations. If MAP/TOP achieves universal acceptance, as appears to be happening, then the "Tower of Babel" problem will be well on its way to resolution.

We are primarily interested in the MAP portion of the MAP/TOP communications protocol standard because it deals with CIM needs. TOP concerns itself with engineering work station standards and things such as office word processors and electronic mail networks. While these are important to the overall CIM solution, they are secondary to the main discussions of this text. Therefore, I will devote my explanation only to MAP.

Figure 11.3, published by the Society of Manufacturing Engineers in 1986, represents the seven-layer communications protocol agreed to by the proposers, the equipment producers, and the technical societies. Version 2.1 has since been superseded by development of more effective communications devices, and no version is ever to be construed to be the final version. However, to prevent massive early retirement of still-useful equipment, the desire is to make devices designed under the newer versions compatible with components designed under older versions. This is necessary for the standards to have any validity over long periods. Therefore, it is sufficient to use the original version for explanatory purposes.

Figure 11.3 shows seven concurrent communications channels. This is thought to be sufficient for even the most complex devices. Not all users need all seven concurrent

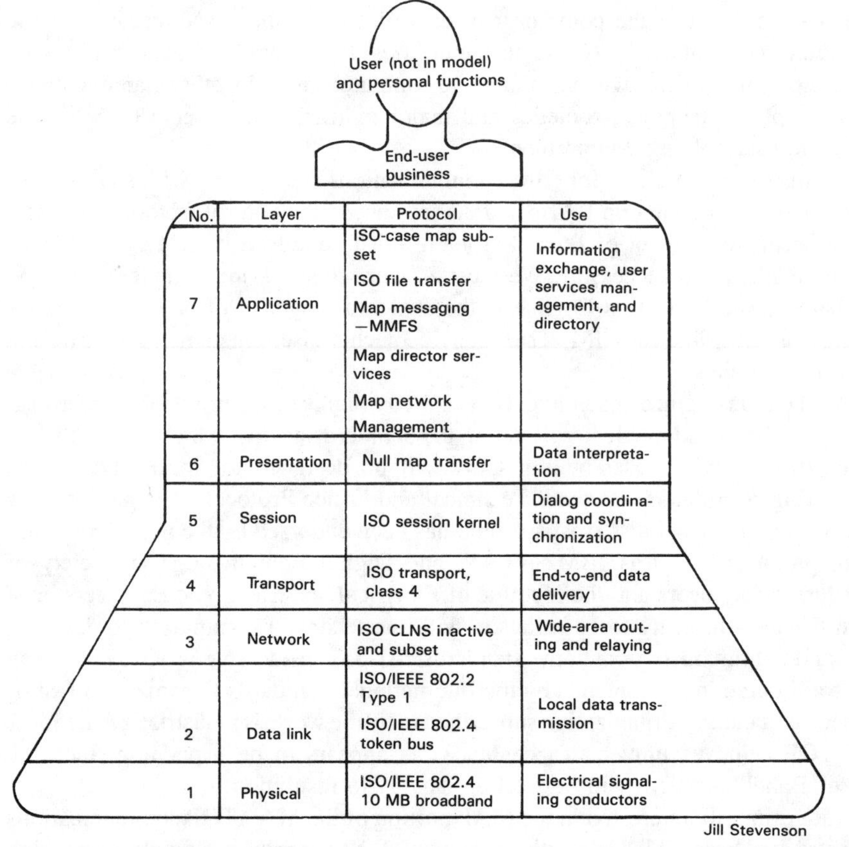

No.	Layer	Protocol	Use
7	Application	ISO-case map subset; ISO file transfer; Map messaging —MMFS; Map director services; Map network Management	Information exchange, user services management, and directory
6	Presentation	Null map transfer	Data interpretation
5	Session	ISO session kernel	Dialog coordination and synchronization
4	Transport	ISO transport, class 4	End-to-end data delivery
3	Network	ISO CLNS inactive and subset	Wide-area routing and relaying
2	Data link	ISO/IEEE 802.2 Type 1; ISO/IEEE 802.4 token bus	Local data transmission
1	Physical	ISO/IEEE 802.4 10 MB broadband	Electrical signaling conductors

Figure 11.3 MAP 2.1 specification and the OSI reference model. (From *CASA Newsletter*, Society of Manufacturing Engineers, Fall 1986, used by permission.)

layers of communication, and users of lesser numbers of levels can just as easily use the protocol. This is like using Lotus 1,2,3. It has many features, some of which may never be used by specific users. OSI means open system interconnection, which is an existing worldwide communications standard. This points out the desire of the proposers to use what already exists to simplify the standardization task. Wherever possible, the MAP specification uses the selection of currently available standard protocols that are suitable for LANs. Where no protocol exists, they have been temporarily filled in with General Motors standards. These interim standards are being replaced with agreed-to standards as they become available. General Motors and all involved with MAP/TOP have been quick to point out that they are not interested in imposing standards, just in setting in motion the development of standards in an expeditious way.

The need for a MAP-type protocol is evident. Most companies recognize the need to coordinate production facilities and automation equipment with operations policies to improve profitability. CIM gives us the methodology to do just that, but it is difficult if all the equipment has to be redesigned to communicate with one another. The establishment of a single protocol, such as MAP, allows all vendors to converge their designs to meet an acceptable standard. Managers recognize that there are many competent vendors supplying CIM equipment. By having a standard, all have an equal chance of winning supplier contracts. It lessens the chance of vendors being squeezed out of the competition because their protocol does not match the predominant protocol existing in the client's plants. Multi-vendor choices based on non-proprietary standards give managers more to choose from. This enhancement of competition will always result in better solutions and thus improved profitability.

This concludes the discussions of networks for CIM communications. It is necessary that the network chosen be truly capable of supporting the intent of the CIM philosophy. We cannot have a successful CIM solution if the cognizant communications procedure is not supportive of integrated solutions. I refer to Figure 11.2, the integrated communication system diagram. CIM needs this. It cannot fulfill its potential without it.

Chapter 12

Managing in a CIM World

What is different? What changes has CIM required of managers to perform effectively in a structure that is dominated by computers and electronic information exchanges? There is plenty that is different and there is plenty that remains the same. The principles of good management, organization, planning, and follow-through remain unchanged. It is the degree and intensity perhaps that have changed, or, to be more precise, computers have offered us the opportunity to become better managers. In this chapter we will explore the nature of the changes in management philosophies and show how CIM offers opportunity to manage better and more productively.

As a point of reference, let us describe how a manager functions without computers and electronic information assists. This is not the ancient world. It is one we knew of perhaps only ten to twenty years ago. Then, the information managers received was from written messages, hand-written and typed (remember carbon copies being just that?), telephone conversations with only a remote possibility of being taped, and lots of verbal communications. From this input, managers had to make decisions, usually based on summaries of summaries of continuously ongoing activities. Managers relied on experiences, hunches, and corroborating data to make decisions. Of course, we still do this, but our data are better and hunches are less important. Managers ten to twenty years ago relied more on estimates of the situation rather than precise knowledge of the current situation, except in those nice circumstances where all pertinent information was readily available and easy to assemble. Management was more freewheeling because it had to be. The manager of the 1960s had the same desire as we do today to gather facts and do analyses before making decisions. However, the capabilities to do so were much less. It was just too difficult to gather precise information on a relatively current basis to make it a requirement for all but the most serious decision making. Complex analyses involving strategies of manufacturing optimizations had to be done manually, taking enormously lengthy periods of time to complete, and even then might have been based on shaky and untested assumptions. The ability to create scenarios and apply "trustworthy" rules of thumb to make quick assessments was more important then than today. Basically, we can say that management of the 1960s and earlier was based on acceptable imprecision.

Contrast the 1960s with what we expect of professionals and managers now (see Figure 12.1). We have available to us in a CIM world vast quantities of information, nicely formatted for any type of analysis we desire. The war cry of CIM is "excellence in communication," and we mean that. Even companies that do not quite measure up to the qualifications to be CIM philosophically driven are dominated by computers, which makes them a step function or two better in the communications arena than their 1960s counterparts. We now demand exact calculations based on the appropriate theory instead of the rule of thumb. Scenario making that the strategic planner employed twenty years ago has evolved from a mass of rules of thumb and empirical calculations to that of a computer model. This computer model drives simulations that have real and current facts as the starting points. Acceptable imprecision based on estimates and hunches is no longer acceptable. If we can have precise solutions for a large segment of our decision making, then why not demand it? We expect that of managers today.

We live in a world of available precision and not acceptable imprecision. For example, engineers trained in the 1950s and 1960s and before were taught how to use a slide rule to do numerous mathematical calculations (multiplying and dividing to trigonometric and logarithmic functions). Virtually everything an engineer had to do mathematically except add and subtract could be done on a slide rule with, for the times, acceptable precision. One of the basic themes student engineers had drilled into

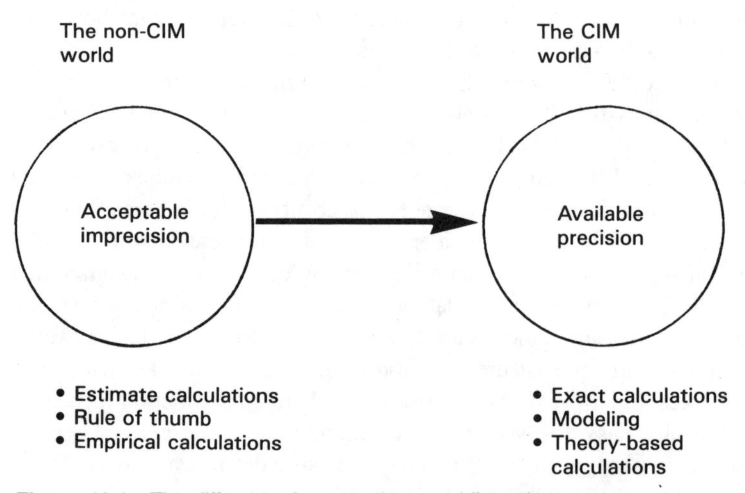

Figure 12.1 The difference in management philosophy

them while learning how to use slide rules was the need to estimate, to estimate to the simplest acceptable level of practical decimal point placement. This was the art of engineering then: make calculations only to the level of accuracy deemed necessary. This was a device to enable the engineer to do his or her job without being overcome by the enormity of risk that a poor estimate could result in. This is why an engineering management ploy of that period was to have more than one, perhaps even more than two engineers, assigned to independently solve the same problem. Then they would get together to check results, similar to checking with classmates after an exam to see if they arrived at the same answer. If the results were similar (they would rarely be the same), then the manager would feel comfortable in using the mean value for his or her solution to the design problem. Today, with the exactness of even hand-held calculators, none of this statistical averaging is necessary. We can easily obtain exact numerical solutions. Thus, the only question that remains is the validity of the data used to obtain the solutions. The electronic hand-held calculator created a revolution in thinking as to how engineers ought to perform their jobs. Acceptable imprecision, along with the slide rule, went the way of the buggy whip into treasured, fondly remembered history. In its place came the theory of available precision.

What does available precision mean to management? It means that there is no excuse for not analyzing a decision to the level of removing all the unknown possible results. We may not be capable of exactly foretelling the outcome, but we should not be surprised by any resultant outcome. With our ability to model, we can create many possible scenarios. These scenarios are all the probable outcomes given the starting and boundary conditions. Now, boundary and starting conditions are not new phenomena. These have been around since Newton invented calculus. It is just that prior to common usage of computers the calculations of solving one scenario were long and tedious enough to preclude any desire to look at many. This is no longer true. Computers do in minutes what the average engineer and manager took hours and days, perhaps weeks, to accomplish. This being the case, available precision a reality, analysis of many scenarios is entirely feasible, and the thrust of management changes from a hunch-derived decision based on experience to that of analysis of various scenarios. We use simulation theory based on our ability to model the situations we are interested in.

Let us now look at two subsets of manufacturing management activities and see how replacement of acceptable imprecision with available precision changes our viewpoint. We will see how CIM has created a new way of looking at the management task. CIM has required a total re-evaluation of how industry is managed.

Quality Control

Acceptable imprecision denotes that it was sufficient to establish acceptable quality levels (AQL) between customer and vendor. Because we could not measure or calculate precisely, we found that clients would negotiate with manufacturers as to how many unacceptable parts are satisfactory for a producer to have in any batch or lot sold to the client. Then, the producer and purchaser would use statistical inspection techniques to determine whether or not the contract was being adhered to. If not, there were financial remedies as well as replacement remedies the buyer could impose. I suppose this is satisfactory in a financial sense, but no one today would accept the concept of acceptable levels of failure.

Today, in the CIM era, only functionally acceptable parts need be included in any production batch or lot. Available precision gives us the capability of making parts several orders of magnitude better than ever before. We commonly deal with parts per million defective as a process control norm, not percentages of the total as in the old AQL system. In fact, it is not unusual to see parts per billion defective as the norm in some aspects of the electronics components industry. CIM, by means of the SQC modules and use of expert systems along with engineering work stations, makes for more precise calculations and monitoring. This results in much higher levels of quality being the expected norm.

With available precision, managers can guarantee virtually perfect quality for all products sold to clients. This is a revolution of thinking brought about by CIM. Available precision now requires managers to be much more precise in their dealings with customers. The expectation levels have risen significantly, which in turn makes managers behave with much more attention to evaluating how they go about their business. Cavalier decisions, some call them creative, can no longer be tolerated. Decisions have to be based on the best possible evaluation of information. The change from the AQL philosophy to the parts per million defective philosophy is certainly responsible for this evolution in business practice. CIM created the ability to have more precise and timely data available and has created a new higher standard to measure management actions against.

Facilities Management

Facilities management has long been the most difficult segment of manufacturing to control. How do you predict when a machine will fail in the factory? Even if you could predict failure, how does one predict how long it will take to repair the non-functioning machine and get it back in operation? These are extremely pressing questions that demand precise answers.

One of the most costly problems in running any manufacturing business is the unknowns created by unplanned down time. In some companies this has led to policies of duplications of facilities and high level of redundancy to act as an insurance policy against unproductive periods. This is an example of the higher and unproductive cost of the acceptable imprecision of the non-CIM world.

With the available precision that CIM fosters, the facility's problem can be solved. Let us look at how the CIM philosophy allows management to solve the failure prediction problem and then be able to accurately predict recovery time.

Mean times between failures are predictable if we understand all the potential causes of failure. This is not a new technology. We have been teaching calculation techniques for mean times to failure in machine design courses for many decades now. The problem associated with accurately predicting anything more complex than fluorescent light tubes' mean time to failure is the complexity of the machinery itself. CIM by means of databases allows us to keep enormous quantities of records. So, it is feasible to input data about failures and by relational databases create any type of comparison we would like to see. If we can find patterns between machine tool usage characteristics and failures, we can predict how long a machine can run before it will fail. These searches can be done quickly and easily. Facilities engineers can evaluate multiple scenarios and thus build models for predicting times between failures. With the access to the database (part of the overall common database system) and use of relational-type software, we are capable of handling the complexity. Before CIM, unraveling the complexities was virtually hopeless. Now, with CIM, it is no longer hopeless. The facilities engineer can do the job because technology has provided him or her with available precision. Knowing the time between failures lets engineers plan for preventive maintenance and accomplish it before the calculated failure would have occurred.

This is really an interim solution, a very powerful and productive interim solution, but still not all that can be done to improve maintenance predictability. We now have a technique known as signature analysis to monitor performance of critical parts of machine tools. This is essentially an ongoing dynamic finite element stress analysis. By locating precision strain gauges on stressed components of the machine, it is possible to literally listen to the machine stretch and compress by means of a technique of calibrated sounds with stress levels. This way it is possible to calculate the number of tension compression cycles the part is undergoing with sound and stress levels, and using fatigue-type equations, to predict very accurately when failure will occur. This is an improvement over the statistical approach just described because we are dealing with the specific reality instead of the statistical mean. In a statistical approach we need to deal with the entire normal curve because we do not know where on the curve the particular machine fits. The statistical approach tells us the time the average machine of this type will run before failure occurs and gives us the standard deviations. From this we can see when the earliest failure will occur and set the time for preventive maintenance to coincide with that. The signature analysis method allows us to monitor the specific machine, which most often means it can run longer producing product for the company before it has to be shut down owing to pending failure. This is simply more economical than the statistical approach.

Once the machine fails, how do we know how long it will take to fix? More important, how do we know how many maintenance personnel and what skills will be necessary to get the machine back into operation? For unplanned maintenance we cannot even guess at the personnel needs. If we use the statistical approach to when to shut down the machine for preventive maintenance, we will have a good idea of what to expect when the machine is torn down. We then can predict the people needs and use the CAPP system to determine how long it will take to fix it. With techniques such as signature analysis, we obtain an even more precise capability to forecast personnel needs because before the machine is taken out of service we will have a detailed description of what will have to be done. This is like surgery before and after the advent of non-invasive testing such as X rays. The signature analysis technique tells us precisely what is wrong with the machine about to enter a preventive maintenance period. With techniques available to us by the various CIM modules, we no longer have to make plans based on hunches. Decisions and planning can be based on the available precision the CIM world offers.

From these two examples, we can see that the CIM world creates a greater capability for the manager to react.

The greater capability gives managers many more and higher quality options to consider when making decisions. Since the options are there, competitive considerations force managers to try to take advantage of them. We can say that CIM requires more precise analysis of options to achieve optimum results. CIM delivers more information to managers, of better quality, and more timely. By the nature of the CIM modules, it enhances the potential for more critical analysis, which, in turn, requires the manager to do a more precise job in using these enhanced resources.

Since the capability is there, companies operating within an implemented (even only partially implemented) CIM environment expect more from their managers. So, we find a strong desire to improve managerial skills, particularly analytical skills. In my experience, I have found that people introduced to a CIM environment as a standard work situation tend to rise to the capabilities offered to them. Operations with lots of computer outlets (terminals, etc.) tend to become more "professional" and more precise in their performance. It is analogous to the theoretical question, which came first, the chicken or the egg. Regardless of the version of the answer preferred, whenever computer equipment is introduced in sufficient quantity to previous non-users, we tend to set in motion an iterative ratcheting up of competency and quality of work output. It is almost as if, because CIM offers more capability, people use it more, thus improving themselves in the process.

While CIM will lead to improved personnel capabilities, it can be a painful process for the individual. Managers have to be aware of the "computer fright syndrome." Imagine the seasoned and successful engineer or manager who is told he or she now needs to use a computer terminal to do his or her work, that no longer can he or she get by with making decisions on expert opinion and experience alone, but now really has to substantiate decisions with analysis and look at alternatives, processes that used to be talked about but rationalized as not necessary because of the pragmatic difficulty of doing it. Now it can be done and the company wants it done. The individual may think he or she is being asked to give up the skills that got him or her to their current position only to find that he or she has to enter competition with a "youngster right out of school," one with a substantial head start in computer mastery. The older engineer or manager begins to question his or her worth and is perhaps bitter that the old way is no longer looked on favorably. This is getting to be a common problem managers have to face. CIM unfortunately accelerates obsolescence of skills.

Managing in the CIM world requires an understanding of human needs, wants, and fears to a much greater degree than before, because we find ourselves in a significant "change" situation. Managers have to be sensitized to this and react accordingly. It is a paradox that the move to the more analytical approach of CIM needs to concern managers more for the emotional well-being of their subordinates than the freewheeling styles of the past. At first look, one would think that is not so. But it is, because the CIM revolution thrives on continuous technological change, requiring rapid and continuous upgrading of skills. It is this state of change being the norm that creates the emotional threats to an individual's sense of worth and, hence, his or her security. We must recognize this potential problem and make plans to deal with it. We simply cannot choose to abandon CIM because of this problem. The potential for CIM is far too great for that to be a valid choice.

What is the basis for the "human acceptance plan" for CIM implementation? It is one of education and making incumbents feel comfortable with the change. The key is much education on all aspects of the implementation plan, to the point of boring redundancy. Along with this, the incumbents have to be made to feel their security is intact and that this change will even enhance that security. Finally, patience becomes a strong management virtue. Unfortunately, we are dealing with years of ingrained methods of doing business and to change that requires the patience of a loving parent. Senior management may be convinced that the old way of doing business is no longer satisfactory, but that is not sufficient. Management has to convince the entire work force that change is necessary and that only through change can the company be successful. Also, company success has to be directly linked to individual success (e.g., security), otherwise the task becomes virtually impossible. One company memo to all employees is not anywhere near enough. The message has to be broadcast at every opportunity, over and over again. This takes enduring patience (see Figure 12.2).

The need for change has to be accepted. This takes convincing of the highest order. It means planning of information meetings where the goals are presented and answers are given in honest ways. This is the time to build teamwork. In fact, one technique that has merit is that whenever change is needed, try to make it occur as a result of good ideas received from the employees themselves. Make it their ideas and their desire to see the change implemented. Get people involved in planning and implementing the changes even though it would be faster if outside experts did the work. Keep in mind that the outsiders may be able to install the hardware and software, but change has not occurred until the workforce is using the new technologies and processes willingly. Above all, to have change accepted, avoid the "we versus they" polarization that too often occurs when the "whiz kids" descend on the organization and run roughshod over everyone in their missionary zeal. Strive for team building and everyone helping each other to accomplish the needed change.

The industrial change brought about by CIM significantly affects management emphasis. We discussed flexible manufacturing systems (FMS) in chapter 5. Now, let us look at how N/C, in its most advanced form, changes management control emphasis.

Traditionally, manufacturing management determined if it were doing an optimum job by basing judgment on the collage of individual operator efficiency measurements. This, of course, was practical and was able to be done when factories were made up of skilled machinists running varieties of manual machine tools. Single operations were done on single purpose machines, and the success of the operations depended on the skills of the individuals controlling those machines. With N/C this all began to change. The operator of an N/C machine does not control the movement of the machine. All he or she does is insert the desired tape into the MCU, load the material, and engage the machine. This means the operator is part of a team, no longer an individual totally responsible for the quality and quantity of parts produced at the work station. This puts the entire efficiency measurement in doubt.

For an FMS, the entire process is automated, from loading the material onto the work table, to selecting the electronic sequence the machine will implement, to

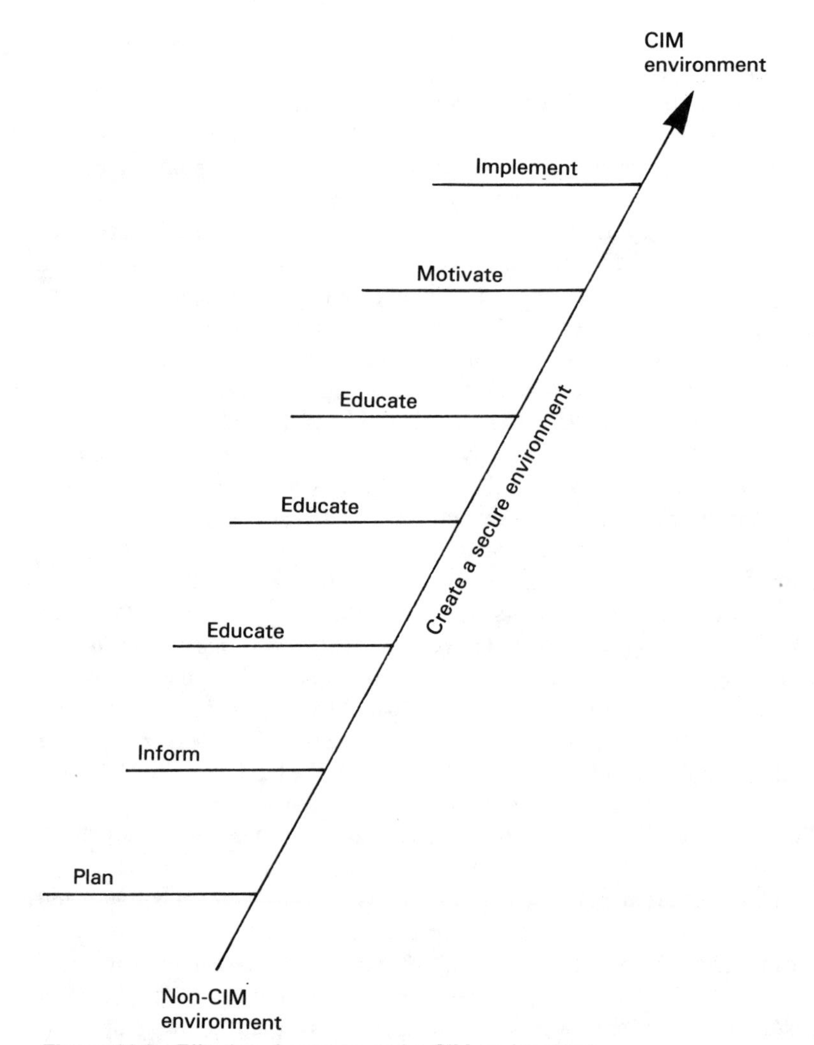

Figure 12.2 Effecting change toward a CIM environment.

performing the multiple machining operations, to inspecting the work performed, and to unloading the completed jobs. The process is pre-planned with the only human role being that of an observer. The concept of measurement cannot be one of operator efficiency because the only thing the operator did was observe and act as a fail-safe device to disengage the machine if something went wrong during the cycle. The only individual measurement possible would be one of omission instead of commission. If the operator failed to turn off the machine if something were askew, then he or she failed to do the job correctly. This is quite a change in emphasis.

In the CIM world we can see that "rugged individualism" of the worker is replaced with "teamwork." So, the measurement becomes how well the group is doing to accomplish the task. We look at how well the supportive communications systems are working and how well the FMS team does in supporting the overall goal of the firm. CIM is creating a greater awareness of the entire functionality of the business. In our example of the FMS, we see a definite trend away from the individual to that of an entire team because the FMS is too complicated to be controlled in its entirety by a single operator. Also, since the FMS is so large a part of the manufacturing capability of the factory versus a single manual machine, the team involved in running it is much more aware of the overall goals of the organization. This has to be since they are responsible for such a large portion of the overall manufacturing output.

The manual machine factory had to depend on the individual operators. If they did their jobs adequately, then the firm would meet its overall goal. If they did not, then the factory failed. So, measuring the individual made sense. The individual operator was not too knowledgeable of the goals of the firm because he or she was such a small part of the congregate total output. This is the reason why operators, when quizzed, only rarely had a realistic picture of the company's status vis-à-vis the balance sheet. From a management viewpoint, this information was not necessary.

The emphasis now is in on responding rapidly to changing business needs, because we can all own FMSs that make fast response easy. So, we see the need for better communication, both technical and business. This is a totally opposite philosophy than the pre-CIM environment.

For these reasons, machine use is no longer a prime measure for manufacturing. The important calibration is how well manufacturing meets marketing's need to have products for it to sell. The key is flexibility and quick response time to the stimuli received from the marketplace. All this leads to more flexible approaches to management. The military-style organization is no longer sufficient and in its place we need a more participative style.

CIM definitely requires a different style of management. We are dealing with communication and ideas as we have never done before. To satisfy this need, we need a much flatter organization matrix with more temporary reporting structures. Let us look at Figure 12.3 and develop what the needs for a CIM world management structure are.

A quick glance at Figure 12.3 would lead us to believe that it is mislabeled. The non-CIM factory appears to be orderly, while the CIM factory appears to be disorderly. This is not the case. In fact, appearance in this instance is misleading. What we have here is a CIM-influenced organization that mirrors the database structures found to be most efficient for handling communication and implementation of decisions.

Recall the discussions of hierarchical versus networking and relational databases. The same premises that made networking and relational database superior to hierarchical databases are true for people organizations too. Rigid structure is necessary when communication is fragile. When the ability to communicate is weak, we need to know where all the players are at all times in order to have any chance at all in getting the message through. The non-CIM factory designates certain functions and sub-functions to carry out specific tasks. These assigned tasks are theirs, regardless of how technology or administrative needs change. They are rigid because the system (or lack of system) cannot cope with change very easily. Since the non-CIM organization does not commu-

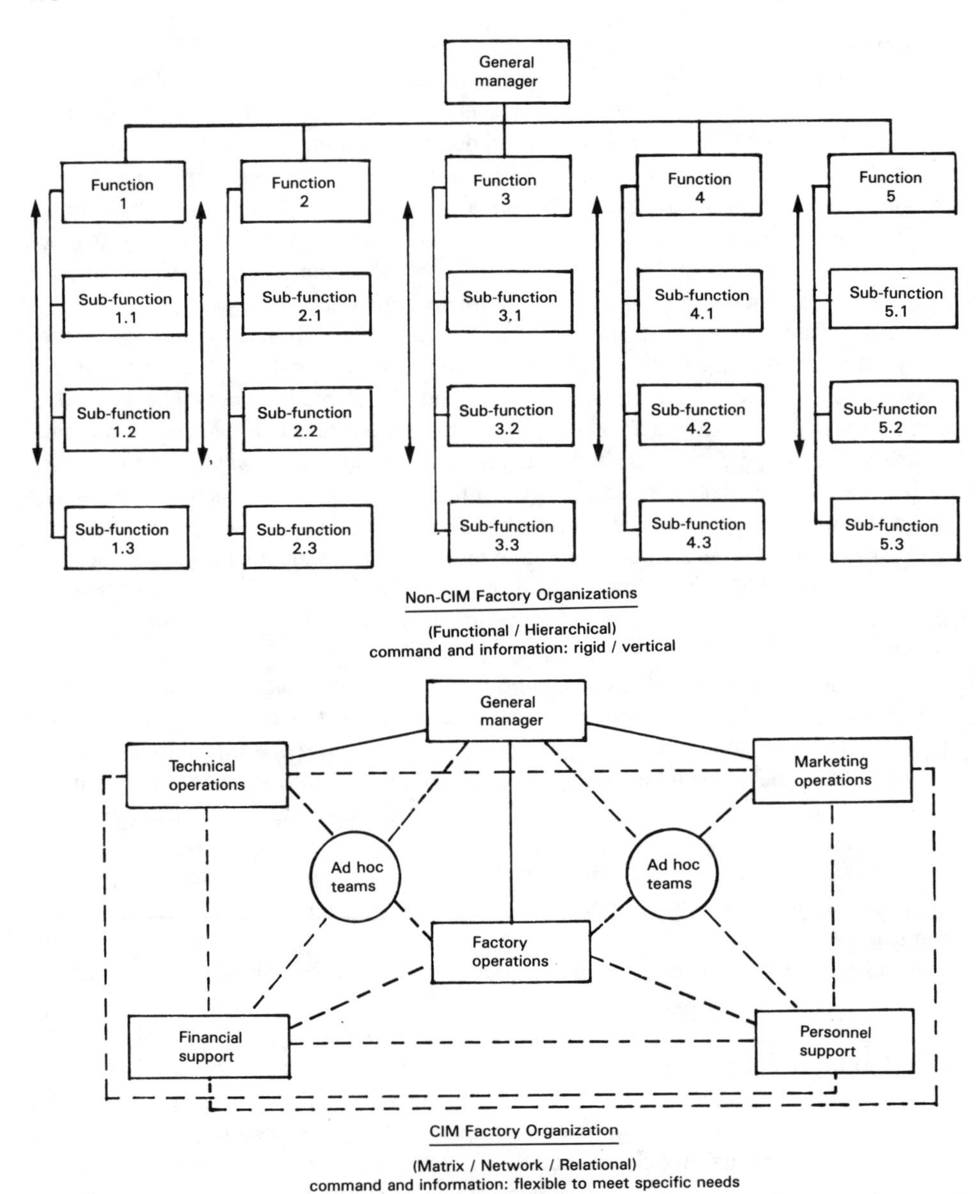

Figure 12.3 A management structure comparison: non-CIM and CIM factories.

nicate effectively, we dare not take any chances and try to expedite by going to a different track to get something done. Let me give an example.

Suppose we need to make a hundred more widgets than the production order called for because of an increased need by our best customer. Also, suppose our best customer wants these hundred extra widgets delivered at the same time as their normal order. Now, the logical thing to do is for the salesperson to tell the shop supervisor to make a hundred more and be done with it. But this is not so logical in a rigid system. If we allow a representative of marketing to tell the supervisor directly to make the extra hundred widgets, we bypass a fragile control structure and will create chaos. The materials people would not know that they should have material for the extra hundred. Nor would they know that the schedule has to be revised for the additional load. Nor would they be aware that the sequence of producing products for other customers had just been delayed. All this occurs because the communication system cannot handle ad hoc change. Therefore, we do not allow any ad hoc activity at all. By doing this, we preserve orderliness and thus control. The fact that our best customer may not get their extra widgets on time to meet their needs does not matter and cannot be helped.

A non-CIM system cannot handle the dynamics of the world marketplace very effectively, because it needs to be rigid to overcome its serious communication ineffectiveness. If it did not do this, not only would it be ineffective, it would be totally impossible to operate. This drawback did not matter much when the whole world was non-CIM. However, that is not the case anymore, and companies that choose to remain non-CIM are at a serious disadvantage. Communication excellence is now mandatory, and, along with that, there needs to be an organization structure that is complementary. Let us look at the CIM factory organization.

The CIM factory organization (see Figure 12.3) takes advantage of the available precision that the CIM world creates. Note that there are only three fixed operational functions reporting to the general manager. One is all the technical activities related to the factory's products. This includes the design of the product and the design of how the product is to be made. A second is factory operations. Here all resources necessary to blend materials, facilities, and labor into product production are controlled. The third is marketing operations, which contain all activities dealing with the external marketplace and its effects on company tactics and strategies. Within these operations we find non-rigid structures similar to those shown in Figure 12.3 in the CIM factory organization portion. The other organizations shown are support components, and they structure themselves to meet the needs of the three operations. This means the support components have to be flexible in their internal structures, similar to their clients.

Now we come to the truly distinguishing difference between the non-CIM and CIM structures. That difference is the recognition of the need for "ad hoc teams." Here we are really taking advantage of available precision. With the precision of a surgeon, we are culling out narrowly defined specialties to solve very specific problems. We form ad hoc teams to do very specific things, then disband them. We can do this because we enjoy communication excellence and the decisions the ad hoc teams make easily integrate into the overall system. If our CIM world factory had to make a hundred extra widgets, we would quickly form an ad hoc team made up of people from marketing and factory operations and do the job. In this example, the team would probably be the salespeople and the shop supervisor. They would not have any problems doing the job without affecting every other order because they would be in communication with all other functions, so changes could be made dynamically and confusion could be avoided.

The ad hoc teams illustrate what is meant by available precision. A decision is made, and it very quickly manifests itself in action. The important thing is that action takes place without disrupting the normal routine. In fact, a case can be made that the normal routine is in reality a series of ad hoc teams addressing here and now challenges. We can see that the CIM factory organization is much more capable of reacting to the dynamism of the world marketplace than its predecessor organization.

One of the most vexing problems facing managers is how to assess the value of CIM by the traditional measurement techniques. If we are going to implement CIM solutions to improve profitability, then we have to be able to measure the worth of CIM is an objective and accurate manner. The non-CIM factory uses worker efficiency and work station efficiency as the primary measurements. We know now that since CIM is an integrating philosophy that emphasizes synergism throughout the entire organization, a piecemeal measurement system satisfactory in the past is not adequate. A measurement system that is fair to CIM and objectively shows the worth of non-CIM solutions is necessary. This has caused new looks at how businesses should measure the worth of their efforts. Let us discuss some of these.

The first we should consider is the productivity ratio. This is an effort to evaluate the worth of an integrated function or set of functions, thus getting away from the drawback of evaluating effectiveness based on individual performance. The old standard way of measuring factory effectiveness was to measure the efficiency of each worker individually and then combine and average the values to come up with an overall assessment of performance. This is fine if all operators really make completed parts (e.g., they start and finish a discrete set of operations individually, resulting in a finished component ready for assembly). The problem, however, is that not all operators are doing the same thing, nor is their individual work of the same value to the company. If it were, they would all be receiving the same rate of pay, which they are not. This means the averages of all the individual operator efficiencies probably is badly skewed one way or the other, giving a distorted appraisal of the situation. When we introduce N/C machines, particularly those capable of doing more than one operation in succession, the problem becomes excessively bad. This is true because we are no longer measuring an operator, but an operator along with an N/C programmer, an MRP II system dispatcher, a maintenance crew, and a material handling system, plus many more support functions. With N/C, we inevitably introduce the CIM team system concept to the measurement task. One way around this is the productivity ratio technique.

The productivity ratio technique compares the amount of time needed to perform a set of tasks the old manual way to the new CIM way. For example, we look at all the operations needed to produce, say, a screwdriver. To produce a screwdriver, we have metal forming, metal shaping, metal machining, and corresponding wood or plastic operations plus assembly. This is a completed component that can be bounded and its manufacturing processes defined. We do not really care about the sequence and the particular operations of how the screwdriver is made as long as we capture all the steps. We then measure portal-to-portal time necessary to complete all the associated processes and divide the old method by the new method to come up with a numerical non-dimensional value. If the value is greater than unity, then the new method is superior to the old. Conversely, if it is less than unity, then it is worse, and at unity there is no difference. This ratio is fair to the old and new method. It is simply an effectiveness evaluation based on cycle time. The premise is that shorter cycle time is better than longer cycle time for efficient use of resources.

The productivity ratio is strictly an effectiveness measurement. It makes no attempt to evaluate the costs of the old and new methods. From that viewpoint, it could be a flawed measurement, particularly if the old method is simple and requires little or no capital. The new method may have to be many orders of magnitude better to justify the cost of the new method.

This leads us to the financial comparisons evaluation of the worth of the CIM solution compared to the non-CIM solution. The financial comparisons usually fall into two categories: direct benefits from using CIM to perform specific tasks and indirect benefits.

Direct benefits are such things as automated drafting systems that usually result in shorter processing times to do the drawing work. The financial comparison is the cost of time saved to do a manual drawing. This is multiplied by the number of drawings performed in a specific time period. This value is then multiplied by the cost of the operator per time period, which gives the savings. The payback period and other cash usage calculations can then be calculated and the CIM module project can be evaluated as any other option would be for capital expenditures.

An example of an indirect benefit of using a CIM solution is N/C tool path simulation, while N/C tool path would be a direct benefit because it directly involves shortening the process cycle time. The simulation involves checking out the path the tool will have to take to make sure it can do the job correctly. This is secondary to making the product and, hence, indirect. Simulation saves money because it eliminates the wasted time of cycling the machine through the steps to make sure it will be adequately positioned for each operation. Usually, N/C machines are meticulously stepped through new programs to make sure they will indeed cut the material in the desired manner. This is done without any material on the machine; and at each major change in operation sequences, measurements are made to determine if, for example, the tool can cut work material, or will it cut the bed of the machine by mistake. All programs are checked because a mistake can be very costly to the health of the machine itself or could result in the work piece being ruined. The calculation of savings does not involve use of the simulation program. It is assumed that the program can be run offline while the machine is running another qualified program. The savings are calculated by determining how many additional productive machine hours will ensue if qualifying a program tape can be done elsewhere.

The major problem in measuring the benefits of the CIM solution is making sure the entire effect of CIM on the overall system is accounted for. We must make sure we combine the tangible and intangible worths (the direct and indirect). We must make sure the limits set for the comparison measurements are broad enough. For example, if we are evaluating an engineering drafting system versus the old manual drawing board and T-square system, then we would be patently unfair if we just considered the increased square footage of drawings the computer-assisted drafting (CAD) system could deliver. These can be very expensive devices, and if square footage of drawings were the only measure, then it is quite conceivable that it would be cheaper to simply take the multiple of the increase and hire an additional draftsperson or two to double or triple the manual output. This would certainly be cheaper. This is an example of the measurement boundaries being too limited. The real benefit of the CAD system is the vastly improved accuracy to be obtained and the fact that it is integrated with the entire common database. Being integrated means that things impossible before become possible. An example of this is N/C tool path generation. There is no reason why the tool path cannot be generated immediately after the completion of the drawing and perhaps at the same time. This eliminates the need for some manufacturing engineering work. This is a benefit that would never be explored if the measurement limits were too narrowly set.

Measurements are but one problem that has to be overcome in managing in the CIM world. Let us look at some others.

Ignorance is an obstacle that has to be overcome. We know that change is not widely accepted in any circumstance. When we are dealing with CIM, we are asking people to change what they axiomatically believed in. They have to be convinced that this major change is not a threat to them. For this reason, many companies have had difficulty convincing senior management to implement CIM. They are not knowledgeable about CIM and perhaps feel that their control is threatened by this amalgam of electronic marvels. To overcome this drawback requires patience, continuous education of senior managers, and demonstrations done enticingly enough to convince them to let it be tried in a pilot fashion. It will require a "champion" with dogged determination to overcome this problem. We will discuss aspects of the solution to this problem in the next two chapters.

Competition for funds to implement CIM by other users of capital have to be met and defeated. Good companies always have many more good uses for capital than capital available. This means the champions of CIM have to do an excellent job in defining the paybacks of CIM projects in a manner that is convincing to the holders of the corporate purse strings. We will cover economic justification in the next chapter.

A third problem involves "keeping up with the Joneses." CIM is still relatively new so that those companies using it successfully really do not say too much about it. It is still considered a very exotic competitive weapon, one that the pioneers expended much sweat and tears to perfect, so they are not about to tell much about what they have learned. This means it is harder than usual to get outside appraisals as to the effectiveness of the CIM modules one might want to implement. It is harder to justify because we cannot as readily use the results of others to aid in the justification. Therefore, information pertaining to the value of CIM tends to be discounted as subjective rather than objective in too many instances.

A truly difficult problem to overcome is the emotional objections to CIM by a significant portion of the union labor movement. There is a fear that CIM is nothing more than automation, which, in their eyes, means job losses. Obviously, they do not understand that this "know nothing" attitude will end up with massive job losses because their companies will not be able to compete with those that do employ CIM. The only way to overcome this myopia is through very patient, very extensive education. We know that companies employing CIM strategies tend to respond to market forces much more rapidly and effectively than those that do not. The CIM companies expand and create more jobs across the board. Those that do not take advantage of new technologies tend to shrink their market share and, in severe cases, disappear.

Finally, a problem very difficult to overcome is incrementalism. This is caused by wishywashy management. They know of CIM, but are not quite sure what to do nor do their instincts tell them to go all out with promising new technology. So they opt for, in their opinion, a rational compromise. They expend their capital on incremental improvement, typically employing technology similar to what is already in place. The compromise is that they will allow a little bit of CIM-like technology to be tried. Usually, small islands of automation occur, and the automating linking of the seven steps of the Manufacturing System never occurs. Since they never get the benefits of "communication excellence," they tend to get tired of the experiment and conclude that CIM is not for their company. How is this overcome? With more education. With planned visits to companies in non-competing industries that have good CIM implementations ongoing, and with superior design of the allowable CIM module installation so a tantalizing taste can be administered. A taste that helps the wishywashy management suck up their guts and say, "Let us try some more CIM."

This concludes the discussion of managing in a CIM world. I think we can safely conclude that management in a system offering the prospect of total and dynamic information availability requires more precision and attention to detail. It requires a higher level of competency at all levels, which means those management skills deemed essential since the beginning of the industrial revolution apply, but perhaps in a more intensive application. In return for this more intense and precise application of management skills, we find that the pot of gold at the end of the rainbow is indeed within our capability to possess.

Chapter 13

Topics on Economic Justification of CIM

Money makes the world go 'round, or so the saying goes. Money definitely is a necessity for making CIM happen. CIM, like any other good idea, requires resources for the ideas to reach practical reality. Up to this point we have spent most of the time on the reason for CIM (i.e., to achieve "communication excellence" so that profitability is optimized). We have discussed the various modules of CIM and how they bolster the ability of a firm to carry out the seven steps of the Manufacturing System. During the course of this discussion, it became obvious that CIM does not come cheap. CIM is a complex series of interrelated technologies joined through common databases and judicious uses of computer sciences blended with engineering reality. This is an expensive proposition, at least for now and for the foreseeable future, but one that promises profit benefits far exceeding the costs. Now, we should discuss the strategies for justifying the costs of CIM and getting the approval to proceed with CIM projects.

In this chapter we will present the basic philosophy of evaluating any capital project, which may be a review for some. Then, we will look at some strategies for making the financial reviews meaningful with respect to obtaining CIM project approvals. Finally, the topic of project management and its relationship to economics will be presented.

CIM has to be justified, just like any other activity a company may consider embarking on. Projects that cost little in cash are justified with respect to the resources the company must supply in order to complete them. We ask, first of all, is this project worth the effort? If it is, then is it worth more than competing (other possible) activities for the resources. As managers, we make these decisions every day in our business careers. When a prospective activity needs cash resources as well as people resources, then we look at the decision making process in a more formal way. We do this because money tends to be harder to come by than people, and because it is easier to measure good and bad applications of money than people. It should not come as any surprise, then, that codified procedures, based on sound reasoning, have been developed to help companies make decisions as to what projects should be supported with funds and which ones should not be.

The first concept we have to consider is the hurdle rate. This has nothing to do with whether or not a certain calculation is used in determining the facts. It is simply the rules as to "safe" or "out" and how the game will be scored. Take baseball: for hardball, the distance between bases is ninety feet, while for softball it is only sixty feet. That means in hardball the runner has to traverse ninety feet to reach the base and in softball, only sixty feet. In both versions, if the runner reaches the base before the baseman receives the ball, then the runner is safe. Therefore, safe or out is measured at the base, not along the path. The same is true in economic justification. Whether a project is accepted or not (safe or out) is judged on its merit. It is the hurdle rate that determines how far the project has to run to reach a decision or evaluation point. The hurdle rate in baseball is 60 or 90 feet before a decision has to be made. In business, the hurdle rate can be a number of things. It can be payback period in number of years before a prospective cash flow is positive. It can be the upper limit to costs that the company feels it can support. It can be a combination of both or something entirely different but pertinent to the company. The bottom line is that a project has to pass an arbitrarily set hurdle rate before the merits will be considered.

Hurdle rates set the stage so management will only spend its time evaluating projects that at least comply with the set business plans for profitability. The equations and scenarios used to evaluate projects just give numerical and intangible results. In itself, the results of such evaluations are meaningless. We need hurdle rates to compare the evaluation results against. For example, it is meaningless to say a goat consumed two quarts of water in twenty-four hours. So what? What is the context of this statement? Unless you happen to be a veterinarian or a goat keeper you have no way of evaluating the importance of that fact. The same is true in business. Typically, a business knows what kind of return on its capital it can justify. So, the hurdle rates are set such that the use of capital will be in accordance with the company's needs. This means the hurdle rates are set, properly so, without any knowledge as to what projects will be evaluated based on them. This ensures that the firm will look at all projects objectively.

No matter how fond we are of CIM, we must keep in mind that the objective of a business is to be profitable. If CIM meets the test, and passes the hurdle rate, then it is fine, and it should be implemented. If it does not pass the hurdle rate, then perhaps the circumstances are not right for CIM, or the case for CIM was not presented and evaluated correctly. In the real world, both situations prevail, but happily the latter far exceeds the former. We can do something about the latter much easier than about the former, and we will discuss that later on. Knowing the hurdle rates, we can establish targets and plans for proposed CIM projects to meet the mark. This way we can ensure that CIM projects receive a fair managerial review.

A hurdle rate is a means of evaluating the results of an economic evaluation.

Economic evaluations take all sorts of shapes, from the very simplest (such as how much does it cost and how much money is in the rainy day fund) to the most complex, taking months to identify every possible scenario on cash flows and interest rates and foreign exchange prognostications. We will only look at the most common economic evaluation techniques.

To make an economic evaluation, there has to be an action plan requiring funding to evaluate. To get any cost-bearing project approved usually requires some degree of skill in preparing what is called "appropriations requests." This, in some ways, is akin to storytelling. There has to be a premise and a means of defending that premise against all things that the storyteller does not want to happen. This means good storytellers weave a convincing tale that the listener finds easy to believe. In the business world this means the story makes good economic sense, and all the logical alternatives are exposed and then methodically destroyed as not being as valid as the preferred solution. This principle holds for CIM projects as well. The "champion" of the project, usually a line manager involved in CIM implementation or the manufacturing engineering manager, prepares a document (the appropriation request) showing the preferred solution and various alternatives associated with the CIM project. These alternatives include other ways of achieving the same or similar results. Some may be other alternatives of the CIM philosophy to be employed, and some may be distinctly non-CIM ways of achieving the limited objectives of the project. The purpose of the appropriation request is to logically shoot down all but the prime method of achieving the results. This is done by using good engineering logic merged with supporting economic logic. If the alternative to be employed is indeed the best choice, this is a relatively easy task.

CIM projects are normally capital intensive. So, it is logical to assume that the ultimate decision as to whether or not to approve the request to proceed will be decided at the highest managerial level of the company. Usually, these high-level managers are not well versed in CIM technology, so they are forced to rely on objective financial analysis as the primary approval or disapproval tool. It is the requesting manager's responsibility to prepare (or have prepared) the best possible financial justification and to back it up with an excellent layperson's description of the technical reasons for the choice in how to solve the problem or to take advantage of the business opportunity. This is not a trivial task. This means the responsible manager has to be cognizant of financial justification concepts and be capable of strategizing their uses to support his or her technical conclusions. The manager must orchestrate the appropriation request such that the desired conclusion prevails.

There are many financial justification procedures available for use. We will only cover, in summary format, the three most commonly used procedures in industry. These are payback period, return on investment, and present worth. Expanded explanations of these three economic evaluation techniques are available in the various engineering economic discussions found in many manufacturing/engineering/management texts, including my previous text, *Manufacturing Engineering: Principles for Optimization* (Hemisphere, 1987).

1. *Payback Period.* The payback period is the simplest financial evaluation and is used primarily for gross estimates on the preliminary stages of a capital funding project. The formula is

$$\text{payback period (years)} = \text{(cost of project)/(incremental project savings per year)}$$

Whether or not the project is deemed satisfactory depends on the hurdle rate set by the company. The hurdle rate-setting philosophy is typically based on the economic circumstances of the company. In times of prosperity, when the company has sufficient capital investment funds available, the hurdle rate may be in the four to five year range. This usually allows projects with large intangible savings to be approved. Unfortunately, many CIM projects fall into this category because information giving and getting improvements are hard to pin down precisely. Companies with tighter economic prospects may lower the hurdle rate to two years or less. Thus, only absolutely sure-bet projects will be considered for approval.

2. *Return on investment.* The second method of evaluating capital funding projects is the return on investment technique (ROI).

$$\text{ROI} = [\text{(average savings per year)} - \text{(cost of project per year of life)}]/\text{cost}$$

This technique takes into account the expected life of the equipment and software or, conversely, the window of time we have to make a profit with the proposed technology implementation. When we stop to think about it, this is a very important concept. If the payback period for, say, a time and attendance data collection system is three years, but our manufacturing engineering and management information systems people tell us that within two years we would need a new and different system to keep on track with the overall CIM implementation plan, then implementing this version of data collection is economically unjustifiable. We would never break even; hence, it is a poor business decision. This would be the same as introducing a new product with a cost structure needing a three-year payback to break even, while marketing can foresee at best a two-year window of sales opportunity.

ROI values are always percentages. In that manner they can be roughly compared to cash investment opportunities. Note in the formula how short life cycle, either equipment or opportunity, unfavorably affects the outcome of the calculation. The formula gives a percentage that we choose to compare with cash investments. However, there is a clearer, more definitive, way of evaluating cash flow opportunities. This leads us to the third method of economic evaluation.

3. *Present worth.* The present worth method uses the classic time value of money formula, where the present value of money (P) equals future value of money (F) divided by 1 plus the interest rate (i) raised to the number of interest rate periods (exponent n). Expressed in notation form,

$$P = F/(1 + i)^n$$

This technique is sometimes called the discounted cash flow (DCF), or internal rate of return (IRR), or net present value (NPV) method. Whatever it is called, the technique is the same. Usually, a target interest is selected, normally the prevailing rate for commercial paper or the prime interest rate. Then the projected savings over a period of years is calculated independently for the proposed CIM or other capital project. This projected savings is deemed the future value of money. The number of years calculated to achieve the savings is the number of interest periods. This is usually the equipment cycle of the product cycle time; for example, the number of years the data collection system is intended to be in use before the payback period reaches break even. The current interest rate is used, and we calculate a present value of the money. If the present value of money comes out to be higher than the cost of the proposed capital investment project, then the project is deemed to be a good project and probably will be approved. The evaluation showed that the company can get more for its money by investing in the proposed project than by investing in the financial markets.

Let us now discuss some very obvious facts about financial evaluation tools. They are only tools to be used by skillful managers. We cannot and must not make CIM project decisions just on the numbers. We must put practical business sense, not banker's logic, into the decision process. There are many investment decisions that transcend our ability to make evaluations based on formulas. For example, a new computer-controlled automated material handling technique may open avenues for complying with safety requirements. This may not be quantifiable; in fact, it may represent a terrible payback. However, to not approve the project may result in government decertification or worse. So here the financial decision based on the numbers available for calculation is secondary. There are many other examples where the intangible benefits are very important and cannot be lost to sight. It is the manager's responsibility to not allow erroneous decisions to be made because too much emphasis is put on the financial aspects of the decision and not enough on the overall well-being and future of the business aspects. CIM definitely has significance concerning the future of the business. CIM represents massive change in how a company operates. Since the changes are massive, quite often the available financial data are just not sufficient to make an easy decision. CIM is the future, and we must help managers make the correct decisions. Sometimes this is not easy; therefore, all "champions" must continue to find ways to make intangible benefits look like tangible benefits and do their best to paint an accurate vision of the CIM future along with a vision of their company's future without CIM.

Intangibles are the bane of CIM justification. Robert S. Kaplan of Harvard University, in his August 1986 presentation on the economic justification in the American Society of Mechanical Engineers (ASME) training videotape, *Computer Integrated Manufacturing Systems Theory and Applications*, has come up with some logical suggestions on how to handle intangibles.

First, he says, totally discounting the worth of intangibles is wrong. He says that arbitrarily establishing a zero value for intangible savings is just as erroneous as establishing any other value. He claims that the proper thing to do is to establish a range of savings that could reasonably be expected to be achieved and use the average for the economic justification procedure. However, he cautions that intangibles should not be used unless the tangible savings are not sufficient to allow the project to meet the hurdle rate. I believe this makes quite a bit of sense. CIM is relatively new. Therefore, it is unreasonable to expect that savings associated with achieving communication excellence are going to be easily defined. We have never before in our economic history been able to propose a technology that offers the feasibility of optimizing communication. So, it is unreasonable to expect that there are firm track records of the financial worth of such systems. However, we do know within ballpark accuracy what such improvements in communication would mean to our specific businesses. These we can estimate, certainly within ranges of expectations. We also have some degree of access to consultants who have worked with various companies on aspects of CIM. Thus, it is possible to obtain information about ranges of savings these other companies have achieved. Equating others' savings with our situation ought to give us a relatively sufficient range of savings we could expect. Therefore, Kaplan's suggestion that a zero worth for intangibles is much too conservative is well taken. We ought not to severely punish our companies by not using a reasonable assumption for savings for intangibles in our economic justification exercises.

The second recommendation Kaplan makes on the ASME tape is that assuming the status quo will prevail over a long period is faulty reasoning. An important part of economic evaluation is to consider the null option (i.e., do nothing). When this is done for use with new technology implementation "go/no-go" decisions, according to Kaplan, the assumption is erroneously made that the cash flow and profitability of the firm will stay the same. This means that the use of the new technology is only a positive addition to the profitability if it is successfully implemented. If the new technology is not used, then the company gains nothing, but neither does it lose anything. Kaplan makes a compelling argument saying this is not true. He argues that to do nothing implies that the competition is also doing nothing, which is probably not true. North American business experiences with Japanese competitors tend to support his argument. If the competition does something with similar technology and implements it, they will improve their profitability. This improvement, then, will be at the expense of the company who did nothing. In fact, their do-nothing scenario is not a null scenario but one where profitability will drop. What Kaplan points out is that excluding the use of intangibles and contentedly believing that the null case is valid can be a pretty big mistake. I believe as a matter of course that when we present management with a CIM project that requires some intangible savings to justify, we must vigorously attack the false security of the null option. If we do not, then we are negligent in carrying out our managerial responsibilities. I suppose makers of buggy whips also felt secure in their reliance on the null option.

As promoters of the better way of doing things, champions of CIM have to continuously strive to find ways to quantify the intangible savings. We can do this by paying attention to how other firms, even competitors, are doing before and after installation of CIM modules. We can rely on old standbys, such as time studies, to gain before and after comparisons and project those results onto other, similar projects. We can also extrapolate savings made in areas such as personnel reductions for engineering automation projects or automated production control projects. These are just a few of the things we can do, and we are limited only by our ability to be creative. I have found that a simple way to get my thought processes going pertaining to intangible saving is to simply list the reasons we want to do a certain project, then force myself to write next to those reasons what I expect to save (keep in mind that a CIM implementation that does not save the company any money is not worth doing), and finally jot down how I think the savings will occur. Once I have my list of how I think the savings will occur, I discuss it with my colleagues and see if we can come up with some way of predicting and measuring those savings. It is amazing how this little bit of straightforward logic reduces a giant iceberg of intangibles down to a few easy-to-live-with ice cubes.

A second phase of financial management skills comes into play once a project is approved. The reason it is necessary to discuss it at the same time as investigating techniques for project justification is because how a project will be managed is vital to whether or not a project will be approved. Think of the senior managers who are making go/no-go decisions on appropriation requests as bankers and the champion of the proposed project as a loan applicant. In that situation, the business person requesting the loan would have to detail how the money is to be spent. The bank has to be assured that the applicant is competent to manage the money and that the proposed use for the money is sound. Therefore, the plans of how a project will be managed often become a part of the financial justification decision-making analysis. For a manager to make a decision as to whether or not to approve a project, there has to be some evidence that the spending of the sum requested will indeed achieve what the project proponents say it will.

Once an appropriation request is approved, the responsible manager will have to carry out the plan for achieving the goals. A big part of this plan will be financial. The manager will find himself or herself with a significant sum of money under his or her control to implement the project. With this money, he or she must design, build, and implement the specifically approved CIM project. This calls for skills in setting up budgets, making the necessary purchases, and paying the bills, all within the pre-established plan. The manager is not starting from the beginning, of course. The basic costs and the prospective schedule had been established for the appropriation request. This becomes the basis for one of the most versatile of all management control charts, the combination budget/network chart.

Figure 13.1 is a budget/network chart for designing and installing an automated stocking and retrieval system (ASRS). Note how all-encompassing the diagram is. It shows how long the project will take, the major milestones to be monitored, and how and in what time frame the manager will have to dispense funds. The funds payment schedule is important to the finance department because it shows when cash must be available. This is critical to the company because it plays an important role in planning when the company will have to have funds available to pay those bills. This is called cash flow management and is vital because few firms fund projects strictly out of retained earnings. Most companies require, out of necessity, that a portion of the funding to finance capital intensive projects such as CIM be loans or revolving lines of credit from their banks. It is schedules for payments as shown on the budget/network charts that go a long way toward helping management make yes or no decisions for capital projects. Many times the payback as measured by any of the three evaluation techniques is satisfactory, but the projects are rejected because the cash flow anticipated cannot be supported by the company. When this happens the champion has to regroup and present a more palatable budget/network chart. With an understanding of the cash flow constraints, the manager has to plan an implementation strategy that is entirely compatible with those constraints. This is a pragmatic reason why this secondary phase of appropriation request planning needs to be included in the request for funds.

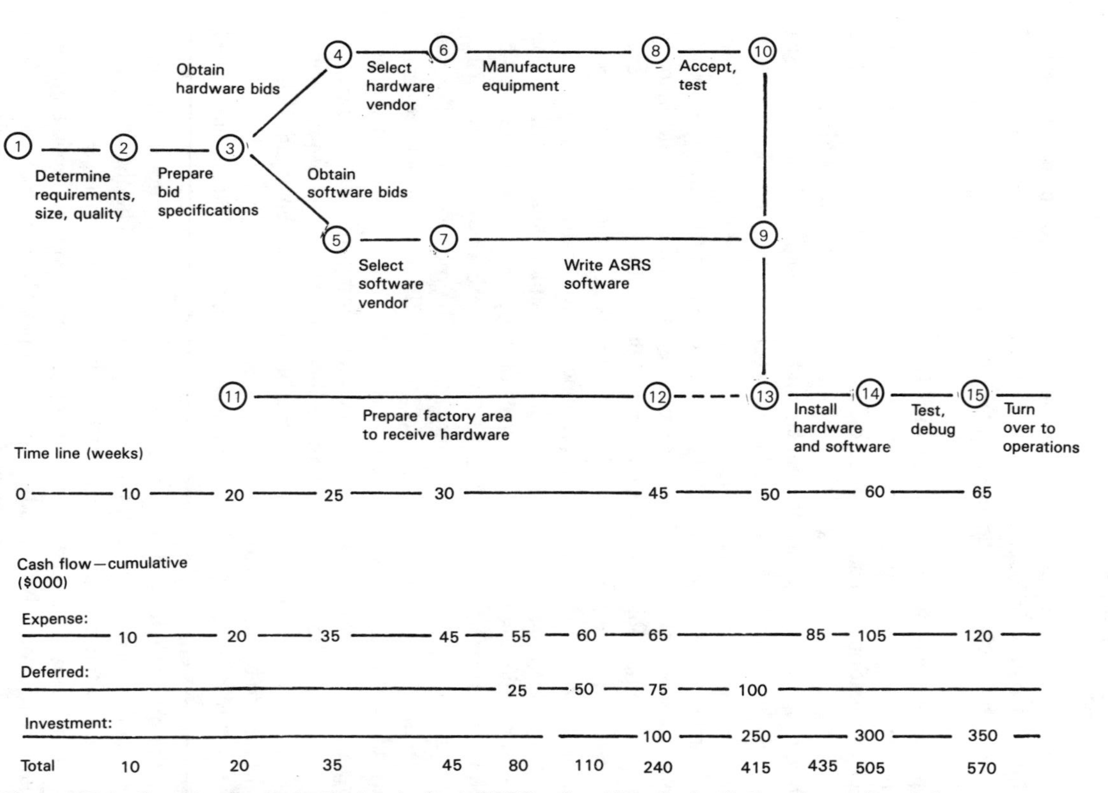

Figure 13.1 Combination budget/network chart. ASRS networking chart with time line and cash flow.

The manager has an obligation to carefully monitor progress and cash flow. He or she must maintain schedule to the best of his or her ability by using appropriate work planning techniques and by above all making sure no funds are paid out that will be contrary to the cash flow dictates. The manager must surely recognize that no payments are to be made before the contractual requirements demand it. This means contractors have to complete specifically agreed-to phases of the project before the payments are made. This is the essence of good management: maintaining control of schedules and assigned funds to achieve desired goals. Including the details of how the manager responsible will be able to carry out these dictates of good management in the appropriation request allows senior management to make judgments as to how realistic the request is and if management appears competent in carrying it out.

Figure 13.1 is a good summary chart for management and is often programmed into an offshoot of the traditional MRP II system for planning and control. The budget/network chart cannot possibly contain all the details of the project, so it is common for it to be the summary for general management, similar to the master production schedule. The details are then cascaded downward to the performing functions to perform in accordance with a detailed schedule similar to the MRP II operations sequencing module. This allows for complete files to be kept of all the specifics, explaining what is required, who will do it, when it will be done, and what the cost will be. This is merged by the common databases with financial records, and a complete audit trail is established.

We now see the overall strategy of writing appropriation requests for CIM projects. We outline the project, its benefits to the company for solving an identified need, the costs, and the savings accrued to the company. We justify the project financially by using recognized financial evaluation techniques. We back up the financial evaluations with detailed project schedules including cash flows to show when funding will be required. With all this we will ensure that the proposal gets a fair hearing, and, if the work is complete and competently done, then we should receive approval to proceed.

That is how it should work, but all too often the financial justification of savings is built on a very shaky foundation. This happens because our ability to quantify cause and effect is not very sharp. We end up making estimates because we do not have sufficient cost data to truly and in detail understand what really happens during the course of running the business. It is this state of affairs that makes us use estimates instead of facts, and, when facts are rare, unfortunately so is appropriation request approval.

This means there needs to be a better understanding of the details of where money is spent and where it can be saved before a successful appropriation request can be written. Ingersoll Engineers, a consulting firm, has pioneered techniques for understanding where money is spent in factories and how to look for savings. Let us take a look at their generally recommended procedures.

Ingersoll Engineers calls their approach "functional costing." They propose this approach to help justify factory expansion (which includes CIM implementation) projects. Recognizing that one of the major stumbling blocks toward gaining appropriation request approval is defining the present costs so that savings can be specifically detailed, the functional costing method aims to correct that situation. The purpose is to help management understand specifically where money is spent in the factory. It is a technique of obtaining specific classification of costs. With this classification of costs, management has a better understanding of the validity of economic evaluations of capital projects.

Ingersoll Engineers has devised three rules for functional costing as shown in Table 13.1. Let us discuss these rules and see how they result in more accurate evaluations of a company's operational costs.

Rule 1: Group costs into four categories:

1. Operator labor
2. Factory overhead
3. Production materials
4. Non-factory costs

If we relate all costs over an entire factory (or product line) in only these major categories, it will yield bottom line numbers that directly relate to the profit and loss statements (P&L). This means we are lining up the categories to directly support development of the P&L statement. If we think about it, it is truly logical. For us to determine if we made a profit, we simply add up all the costs incurred before the customer gives us his or her check to purchase the product. If the check is higher than the costs incurred, then we have made a profit. The problem arises in knowing how to categorize the subsets of activities so we know what components of the company operated efficiently and what ones did not. If we knew that we could fix the ones that need fixing, and leave alone those that are satisfactory. Current accounting practices tend to confuse this simple concept because they have to be tuned to the vagaries of the tax laws, and if anyone thinks tax laws are logical in a true P&L sense, then they are suffering from myopia. Rule 1 tries to correct this situation. We know that making a product requires these basics: the actual hands-on activity to fabricate and assemble the product (operator labor); the design of the product and the process to make the product and the upkeep of the equipment (factory overhead); the purchase and maintenance of the materials to be used in making the product (production material); the sales and advertising, the distribution, the legal, and the other support function activities; and the maintenance of good will with potential customers (non-factory costs). Any cost incurred ought to be counted under one of these four categories. If we do, then we

Table 13.1 Functional Costing Rules

Rule 1.	Group costs into four categories: • Operator labor • Factory overhead • Production material • Non-factory costs
Rule 2.	Group costs together if they are a result of a single decision.
Rule 3.	Assign meaningful (specific) names to all line items.

Note: Adapted from J. Stanton McGroarty, Ingersoll Engineers, "Functional costing: understanding where the money goes," *Manufacturing Engineering*, Jan. 1987. Used by permission.

are really capturing costs in accordance with the way products are made. Therefore, any capital improvement projects (including CIM), if they are true productivity (hence, profitability) enhancements, will affect these costs in a positive manner. If the costs are highlighted, then an appropriation request should be able to show in a cause-and-effect way how the proposed project will eliminate or reduce those specific costs. The cause and effect may still contain intangibles or difficult-to-define relationships may still be difficult to understand, but at least we are linking a process directly with a targeted cost reduction. Making the relationship with the four categories is the only practical way of ensuring that all costs are covered.

Rule 2: Group costs together if they are a result of a single decision. All too often we try to fit costs into a preconceived financial algorithm meant for tax purposes or some other esoteric requirement. Thus, for example, we make a decision to buy a new grade of cutting fluid, which on the surface appears to save the company an attractive amount of money. That fluid requires different types of tool inserts for certain machine tools. The new tools will require different methods to be developed and employed. The cost of the new cutting fluid, the new tool inserts, and the development and training associated with the new method must be lumped together to get the true cost to compare with the true savings. This grouping of costs as a result of a single decision allows us to see if this is a worthwhile project.

Another place where we instinctively separate costs when perhaps they should not always be is between fixed and variable costs. For example, factory supervision may at times be classified as variable, particularly if the supervisor is correcting a quality problem not anticipated at budget formulation time. Therefore, if we have a project that is to reduce fixed factory supervision costs, we may be going ahead with it oblivious to the variable costs associated with total supervision costs. Many companies find themselves in dire straits because they have cut personnel costs too much, and it is usually as a result of not properly taking into account the so-called variable costs. Grouping of all costs associated with a decision is one way of eliminating such fiascos. With respect to the fixed and variable distinction, managers would do well to remember that this is an artificial distinction. Every cost is variable until incurred; thereafter, it is fixed.

Rule 3: Assign meaningful names to all line items. Do not use titles like "incidental costs, other." This does not mean a thing. We can fit everything from the purchase of a can of shoe polish to the cost of site surveys for a new factory into such a broad category. If we are to have an ability to relate costs to potential savings derived from capital intensive projects, we must be able to define all the costs. If the "incidental costs, other" is really shoe polish, then it should be so listed. This allows us to ask the question, why did we buy shoe polish? The answers can be very surprising. For example, if we are a furniture maker, we may find out that because we do a poor finishing job on the cabinets, our shipping department is touching up with shoe polish prior to loading them onto the trailer trucks. Here, the accounting procedure may be hiding a real problem. The shippers are doing what they can to make the product presentable. They are attacking a symptom, not the real problem. They are buying shoe polish, of all things, to get out of a bind they are in, while the cause of the problem goes unsolved. We are not addressing the real problem of poor finishing.

Meaningful titles for accounts allow us to make meaningful assessments of the levels of cost. If the shoe polish example shows us anything, it shows that it triggers a need for management to respond to. Either the shipping department is going into the bootblack business as an aside or there is a problem to be resolved. This example may be silly, but it points out a truism: Ask the right questions, and the most complex problems can be solved; ask the wrong questions, and no matter how simple the problem solution may be, it will not be solved.

With these three rules we can establish a methodology of categorizing costs in a way that has meaning. It is possible to pinpoint improvement opportunities that can be achieved through proper project initiation. Let me give another example. If we break down the cost of material controls to facilities, purchase prices, and people, then we may find that our purchases are costing too much because we cannot pay our bills soon enough to take advantage of early payment discounts. We also may find that the number of people involved in materials is excessive. We see that there are a large number of individuals listed under the title of "production expeditors." We see these facts, and now we want to do something about it. We have dollar costs associated with each of these line items. What we have goes a long way toward justification of an MRP II implementation. If we could not break down the materials costs into meaningful categories, all we would have are guesses as to the reasons. Even if we did a study to break down costs for a one-shot investigation, we may still not know enough to make a decision as to what to do because we may not have broken down the costs sufficiently. Besides, even if we did, then we would be unable to sustain it to prove that the proposed fix actually solved the problem. This would make our budget/network chart difficult to construct and to sell to senior management. Therefore, we can see that the functional costing method has validity and is a powerful tool for managers to have in their arsenal.

Ingersoll Engineers also recommends that where possible show all costs by line item in three categories: annual costs, product cost in percentage of the entire factory's total, and in cost per unit. These three categories are selected because they represent the basic way of thinking about costs for appropriation request decision-making procedures. We look at annual costs because we like to measure ourselves in fiscal year periods. This accommodates tax requirements and is a simple yardstick for measuring the effectiveness of the organization. Product cost as a percentage of total simply allows us to judge the effect of the cash outlay against savings for the product line and its overall effect on the business. Remember, the goal is always to maximize funds usage to make the company more and more profitable over time. Finally, cost per unit is a simple but effective way to measure trends in overall efficiency. If a project is truly worthwhile, then it will reduce the cost of each unit to be produced. Care is needed here to make sure we are measuring all costs so that this is a real indicator of progress. Functional costing helps us do that.

We should also list all costs in descending order of importance, the highest cost to the lowest cost categories. This is a psychological trick that helps focus management attention on the most important areas first.

Putting costs together in accordance with the functional costing rules is an effective method for comparing before-and-after scenarios. It is a way of pointing out where the problem areas are, so action (including the CIM implementation scenario) can be contemplated on the basis of facts. These action scenarios evolve into well-documented appropriation requests that tell a logical story backed up with facts. Using these rules goes a long way toward minimizing supposition and conjecture in reaching decisions on CIM projects.

So far, we have discussed justification philosophies for individual projects. That would be sufficient if CIM were one project. Unfortunately, it is not. CIM is a series of modules linked together through the common databases. CIM is a philosophy dedicated to achieving optimum results through communication excellence by computer-enhanced technologies. This means that to implement CIM we have to have a whole stream of related projects in accordance with a master plan. This will require a continuum of projects needing appropriation requests and going through a rigorous approval routine. The activity of getting a continuum approved can be long and perilous if it is not handled in a strategic manner. The strategy involves linking project implementation plans with economics. We know that capital projects gain approval primarily on their economic benefits to the company. We also know that CIM benefits sometimes defy easy benefit associations because of the institutionalized structures of business. Our discussions up to now were about how to structure appropriation requests to gain acceptance, but sometimes no matter what we do an individual project will not pass the hurdle rate. If that project is an important cog in making CIM happen, then other ways have to be found to get approval to proceed.

Let us now discuss how we can get entire programs past the hurdle rates and ensure we get a shot at successfully implementing CIM. The next chapter will deal with project management strategies to give us successful CIM implementation once the funding hurdle rates have been overcome. But now we have to deal with the process of getting the entire continuum over the economic justification hurdle. We will make the assumption that the individual projects are prepared in accordance with advice given earlier in this chapter.

Typically, most companies adopt a "pay as you go" rationale for implementing CIM. They do this because they have no real experience with integrated uses of computer technologies and are not quite sure they believe in the prospective benefits. However, most companies are willing to give it a fair try if it does not cost too much, particularly if there is a strong champion espousing its merits. What this all means is that the first project of the CIM continuum had better be successful and achieve a handsome payback. Otherwise, it will be difficult to get approval to do project number two. The financial return on the first project becomes, whether we like it or not, the seed money for the succeeding projects.

The cascading effect means we have to use strategy properly to design the sequence of project implementation. In practice, this means going to senior management with a list of all the projects that have to be accomplished to achieve CIM. It also means telling senior management how much we estimate the entire continuum will cost and what the savings will be in total. We then proceed to justify the first project and accomplish it. With an accomplishment under our belt, we now specifically justify the second project. Accomplish project number two and ask for permission to do project number three and so forth, until the entire continuum is completed. At this stage, CIM is fully implemented and achieving the success we forecast. That is the theory. In the real world, how do we do it?

Table 13.2 lists the steps in accomplishing financial approval for a continuum of CIM projects. Let us look at those steps and understand them.

First we acquiesce to the "pay as you go" principle (there's no sense tilting at windmills to fight a corporate attitude that will waste the champion's energy to overcome). Then, we look for a "sure bet" project to be the first. We find a CIM project that is well bounded and has a direct cause-and-effect relationship. This way we know that the savings can be generated and directly attributed to this project. The appropriation request is drafted, and the approval process is complied with. As part of that approval process we make sure all succeeding projects are defined in a manner that will link the success of the "sure bet" with future projects. This makes future approvals easier to obtain. One caution here, though: the highest payback project is not necessarily the "sure bet" we are looking for. The highest payback project may be too complex to succeed on the first try for accomplishing a CIM module implementation. It is more important to have a unqualified success than a big payback for a firm's first attempt at implementing a CIM module.

Another piece of advice concerning project number one. If at all possible it ought to involve technology that has many applications throughout the company. This leads to the second consideration of the strategy. To demonstrate a technology that has potential in many areas of the company, do a pilot project. A pilot project is a smallish project that, although it saves money for the company, its main purpose is to demonstrate technical and financial feasibility. It is always easier to scale up after a success than scale down if the project is only partially successful. Also, pilot projects are psychologically thought of as experiments. Experiments are not always successful, so there is much less trauma associated with the desire to fully implement CIM if the pilot project fails. Pilot projects greatly minimize risks while they make it easier for senior management to approve. The funding is relatively small, so the gamble is minimized. If it proves successful (as the champion knows his or her "sure bet" will), then it will be easier to get additional funding to expand to other areas.

Wherever possible, sell the concept of infrastructure. This is the third point of the strategy. Sell the benefits of infrastructure in a positive manner. Keep philosophizing

Table 13.2 Relating Project Implementation Schedules to Economics

1. Use "pay-as-you-go" process.
   a. Projects (modules) have to demonstrate a financial payback.
   b. Return on one project is seed money for the next.
      (1) Highest payback first.
      (2) Favor simple versus complex projects for priority.
2. Demonstrate technology through low-risk pilot projects.
   a. Funds easier to obtain.
   b. Consequence of failure less.
3. Stress concept of infrastructure projects over stand-alone projects.
4. Do politically favored projects as soon as practical.

that the benefits of CIM are more than the sum of the parts. Keep hinting that the benefits of having the entire CIM philosophy implemented will yield geometric progression savings, not just arithmetic progression savings. But be careful not to oversell. Introduce concepts of networking, databases, the seven steps of the Manufacturing System, and FMS. Do this by regularly scheduled briefings on new technologies for senior management. A successful champion finds ways to bring the message home to senior management in many ways beyond the presentation of the appropriation request. Wherever possible, the appropriation request is pre-sold by having invested in education for the senior managers involved in the approval process.

Stressing the need for infrastructure also lessens the risk of failure. It gives backers more than one chance for success. For example, suppose CAPP implementation gains only limited financial success in the area of reducing cycle times. When CAPP is added to the next project, say, MRP II, we find that it allows it to be very successful owing to the fact that CAPP is a necessary ingredient for optimum MRP II implementation. This, in turn, means CAPP is more successful because reduced cycle time takes on more meaning. We no longer are confined to CAPP savings being the number of engineering hours saved but being expanded to hours saved for the entire manufacturing cycle. Here we have had more than one chance for CAPP to be successful. This is an example of piggybacking infrastructure projects. Stressing infrastructure allows us to minimize the danger of implementing "islands of automation." Thinking in CIM concepts stresses synergism, which we want to achieve.

Finally, be politically astute. If senior management wants to do particular aspects of CIM implementation first, then try to be accommodating. This way we transfer ownership of CIM to where we wanted it in the first place, the senior management level. We have created a very high-level, effective champion. If this is the case, try very hard to do implementation as the new champion wants it. The downside of this is that the proposers have to be careful that they still only do what is possible. This still means success has to be achieved, and we have to cascade projects one at a time, building on previous successes. We have to be honest with senior management if their desires will not lead to success.

This concludes our discussion of economic justification. We must keep in mind that CIM is only useful if it leads to improved profitability. CIM also has to compete with other uses of company funds. So, like excellent advocates, we the champions of CIM must do an outstanding job in presenting our economic and strategic justifications. To do less just might betray the future viability of our companies.

Chapter 14

Strategies for Implementing CIM

When we consider using the CIM philosophy to optimize the procedures for managing our businesses, we are really considering the need to radically change the way we conduct business. Chapter 12 deals with the needs for making changes in our management outlook to effectively implement a CIM system. In this chapter we will discuss how to go about making those changes. We know that implementing CIM is, in the long term, very beneficial and may in fact be of dire necessity for a company's survival. But how do you undertake such a monumental task? Is it possible to really effect such massive change and still survive? In this chapter I will give my views on both of these questions.

Implementing CIM can be done, and it is certainly desirable. The problem is in structuring a game plan that will work for the specific company. CIM is an overall concept for dealing with the dynamic nature of modern manufacturing businesses using the computer as an enhancer, a tool. The basis is the common databases used to achieve communication excellence. Therefore, it should come as no surprise that the implementation of a CIM philosophy within a company is essentially the task of developing the common databases and their associated inputs and outputs. In this chapter we will go through the sequential steps and side plays necessary to achieve a CIM solution. It will be primarily concerned with the development of the common databases tailored to the specific firm. I will describe the process in much detail, and, if I do my job effectively, leave you, the reader, with a decent plan of attack that can be adapted for your own specific needs.

When we talk about CIM being compatible with the theory of manufacturing (see

Table 2.1), we mean computerizing the seven steps of the Manufacturing System in a way that provides linkage between the steps. In other words, we strive to create and maintain a continuous system where the outputs of the preceding step form the basis of the inputs for the current step, and so forth. This is the essence of the strategy for implementing CIM: work with and understand the seven steps of the manufacturing system and establish a plan to replace manual control activities with computerized linked segments. This way we develop an integrated computer-based manufacturing system (ICBMS). It is this ICBMS that is the optimized system we are trying to achieve.

Table 14.1 lists the implementation tasks necessary to accomplish (in sequential order) to achieve a CIM-based company. If these tasks are done faithfully and completely, the firm will develop an integrated computer-based manufacturing system that is efficient and capable of achieving huge profitability dividends.

To accomplish the required tasks, a control mechanism has to be created. I call it the "systems council." This council is made up of decision makers from all operational facets of the business. Usually management information systems (MIS) people are not members of the systems council for good reason. It is important for the users of the future CIM system to be the designers of its required attributes. This way we gain ownership of the process and a commitment to achieving success. MIS people are not users in an operational sense; hence, their viewpoint is different and perhaps contrary to the real operational needs. For this reason, MIS people are not recommended as members of the systems council. However, their technical resources should be employed by the systems council in whatever advisory role deemed necessary.

Systems council members are typically mid-level to senior managers with a good understanding of how their portions of the business function. They also, by necessity, have to be dedicated to the improvement and survival of the firm and not satisfied with the status quo. They are definitely not individuals who have been put out to pasture waiting for retirement, nor can they be timid and afraid to face controversy. They will

Table 14.1 The Task Sequence for Implementing CIM

1. Understand the current manufacturing flow procedures, processes, and information system.
2. By using good industrial engineering techniques, evaluate each process and procedure for opportunities for improvement in conjunction with the seven steps of the Manufacturing System.
3. Develop an ideal Manufacturing System flow for the particular factory.
4. Test the ideal Manufacturing System flow against existing specific business constraints.
5. On the basis of constraints, develop an optimized Manufacturing System flow.
6. Select the order of implementation of subsets of the optimized Manufacturing System flow based on company needs and financial and technical resources constraints.
7. Use the prototype process with vendors to modify their existing software offerings to comply with the optimized Manufacturing System flow, and create an acceptable integrated computer-based manufacturing system.
8. Review vendor offerings of the integrated computer-based Manufacturing System with the user community. Modify as required, and gain user acceptance.
9. Use good project management techniques to implement the integrated computer-based Manufacturing System.
10. Implement automated process equipment, that is compatible with the integrated computer-based Manufacturing System, only after the basic system is operational.

be investigating how the company conducts itself now, and they will have to question everything. They may have to excise many sacred cows to get to the truth and implement needed change. They will not be popular with the bureaucracy, but they will gain an enormous sense of satisfaction once the job is done. If company history is recorded, these will be the individuals who will be noted as those who saved the company through their insight and courage to make major changes.

The systems council will have a simple charge. They will study the current ways their company achieves the seven steps of the Manufacturing System. They then will look for ways to computerize and link the steps in an optimal fashion. They are to create a blueprint for implementation of the CIM modules, ultimately leading to a totally integrated computer-based manufacturing system. The goal of the systems council is to create a design concept for the ICBMS with a summarized implementation and cost schedule. The goal does not include writing appropriation requests. The actual writing of appropriation requests and the supporting justification techniques will be done by the proper organizations in accordance with the techniques described in chapter 13. The systems council is responsible for providing the vision and the guidance for the company's transformation into the CIM age.

The ten tasks to accomplish this transformation are listed in Table 14.1. The systems council is responsible for having these tasks carried out effectively. Let us look at these tasks individually and explain the nature of each. The ten tasks are divided into two basic segments. The first five tasks are self examination and conceptualization of the changes. The last five tasks deal with the real-world job of creating a plan to translate the agreed-to concept into practical reality. So, we see a dichotomy that has to be bridged. The systems council first has to be blue sky idealists, dreamers; then, they have to firmly plant their feet and develop a pragmatic plan that can be done within the current or very near-term future state-of-the-art computer technologies. The major difficulty facing the systems council will be how to shift gears and to make sure that expediency never rules.

1. *Understand the current manufacturing flow procedures, processes, and information system.* Before we can make changes, we have to understand the current way of doing things. This is used as the benchmark or reference point against which all recommendations for change will be measured.

By using techniques of flow charting, the systems council traces jobs through the factory from order entry to design specification, scheduling and purchasing materials, actual factory processing, and all the way to shipment to the customer. Here we ask the six questions: What? Where? When? How? Who? Each followed by, why? The systems council must understand the details of how their company makes its products. They must understand down to the level of those who actually perform the various communication and control jobs. They have to create the base from which all improvements can be made so that a CIM system can be designed and tested. The output of this investigation will be a horn of plenty containing just about every type of manual, semi-automated, and standalone computer system imaginable. It will be a maze. It will take patience and dedication to sort through it and test it for validity. But, at the end, the systems council will have an excellent understanding of how the company is currently achieving the seven steps of the Manufacturing System.

This sounds like a difficult task, and it is; but, it is not impossible. Remember, the task is not to understand how the MPU of an N/C machine works, but where it gets its information from and what the communication outputs are. For example, an N/C machine receives tapes from the dispatcher, and it receives materials from the inventory control/storeroom people. The sequence of when the parts are to be made comes from a schedule sent to the work station by the dispatcher. Once the job is completed, the operator or supervisor reports this fact to the dispatcher, and so forth. Step 1 requires that this information flow in the way it actually happens so we can analyze it for possible improvements. We are not interested in details as to how the N/C machine tapes are programmed, but we are very interested in how the programmer is told to create the specific tape for the specific job. Where does he or she get the request from, and how does he or she respond to it? This is the type of information needed to understand the communication flow. We are in need of information to understand the communication sequencing and the types of information (data) transmitted back and forth. We need to know this so we can construct a database that will receive and transmit this information to and from the various users in accordance with some sort of scheduling algorithm. We are striving to construct a system of communication excellence.

How do we go about collecting this sort of information and recording it in a manner that is decipherable to the entire systems council? The system I have used over the years is called the "brown paper" method. Its inventor, unfortunately, is unknown, lost to antiquity. It is a tool used by many consultants to understand what they are dealing with when they are asked to come to a company to help solve problems. It is a way to get smart about a problem in a hurry. Therefore, it is an ideal tool for the systems council. The brown paper is simply that. It is a roll of brown paper, three to four feet wide by as many yards long as are needed to complete a specific flow chart. It is probably the same type of paper found in vintage butcher shops. It does not have to be brown, but it most often comes from the paper company in that color.

The process is to list each and every communications activity associated with doing a specific task, say, order entry, in sequence. The inputs, whatever they are, are shown as descriptive data blocks with output lines to the next activity. In other words, the output of one activity is the input of the next activity and so forth, until we come to the end of the particular subset we are creating a flow chart for. We use every means possible to show how the information is transmitted, stored, received, and acted on. This means if we normally receive requests for quote (RFQ) as the starting document for the order entry trail, we will actually paste a blank RFQ onto the brown paper in the correct location. This may sound silly, but it is not. It is not silly because that printed form will contain specifics about how the information is received, and, if we eventually want to computerize it (e.g., devise a computer input screen for access to the common database), then we are going to need to know the exact details that have to be entered. Every transaction is recorded on the brown paper the way it actually happens. If it, for example, is verbal by telephone, then we will paste a cutout of a telephone on the brown paper and list what information is given and received. I think by now the word picture is complete. A brown paper is an excruciatingly detailed flow chart of how we communicate and command in order to carry out the specific business function.

Figure 14.1 is an example of a segment from a typical brown paper. The brown paper requires exact information. How do we get this? We do it by accomplishing the following goals:

- Interview the people who actually do the work.
- Observe the work being performed.
- Gather pertinent documents used in doing the work.
- Put the flow, the information and documents, in sequential order on the brown paper.
- Review the flow with the people interviewed.
- Correct mistakes, as required.

The time necessary to do a brown paper varies. It is hard to say when the job is done except to say that it is an iterative process that will involve the person doing the brown paper and those supplying the information. Sometimes those asked for information may be reluctant to give it out of fear that the results will get them in trouble. Obviously, management must do all it can to alleviate such fears, but sometimes it cannot, especially in larger organizations where the bureaucracy is well entrenched. This has caused delays in completing brown papers, and knowing it can happen requires patience on the part of the interviewer plus care to make sure all the questions are asked from as many different viewpoints as possible. This, plus the observations, usually ensures that the information is accurate.

Before starting out on the brown paper trail, the systems council should scope the extent of the effort and determine who will do the job. Ideally, the members of the council should do the job. If that is impractical owing to the size of the task, or other job requirements (although systems council members must consider this task as a primary task and be willing to devote up to 50% of their time to it), then others will have to be given the task. If a systems council member cannot do the brown paper, then a suitable substitute is usually the immediate manager whose area is being flow charted. This individual will undoubtably be affected by the changes to occur; therefore, it is reasonable to get that person involved. Sometimes, a combination of the immediate manager and the cognizant systems council member working as a team is satisfactory. Whoever is charged with doing the brown paper, it is the systems council's responsibility to see that it is done correctly. The brown paper forms the basis for all future improvements and the character of the evolved CIM plan. It is essential that it be correct.

2. *By using good industrial engineering techniques, evaluate each process and procedure for opportunities for improvement in conjunction with the seven steps of the Manufacturing System.* Task 2 is a review of the facts brought out by the brown paper exercise to determine opportunities for improvement. Here we review the process outlined by the brown paper and compare it with the seven steps of the Manufacturing System. The systems council determines which of the seven steps are accomplished in accordance with predetermined documented procedures and which with ad hoc, spur-of-the-moment logic. They look hard and in detail at how the seven steps are accomplished.

What do we mean by "good industrial engineering practices?" We mean that

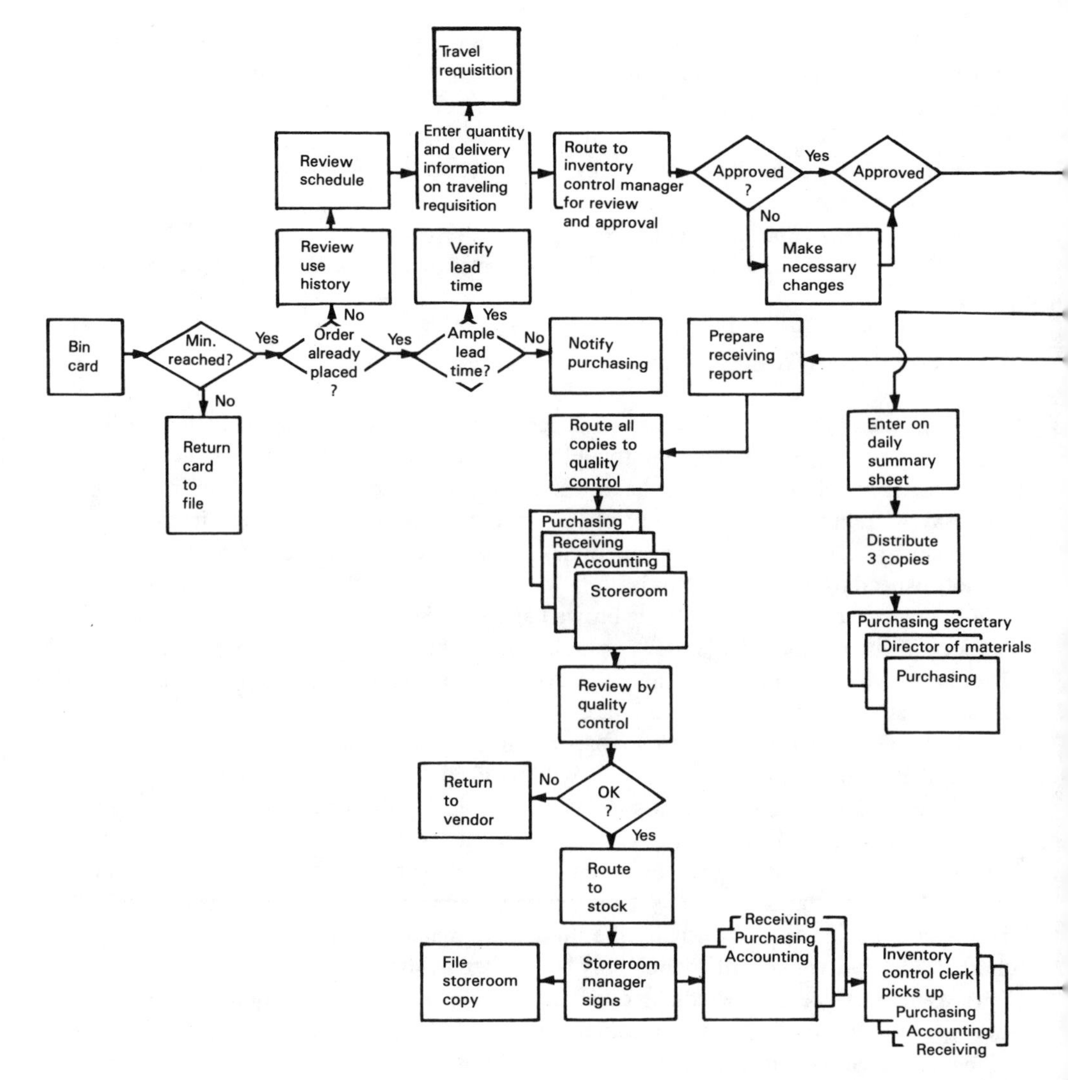

Figure 14.1 Examp

the systems council uses principles of queuing theory, flow charting, and operations sequencing, to name a few of the more common techniques used to analyze information presented in flow format. Perhaps regression analysis and linear programming will also be required, along with statistical algorithms and operations research methodology. What this means is that all aspects of industrial engineering technologies can and should be brought to use, as required, to gain a thorough understanding of the data available and to reach valid conclusions. This does not mean that all members of the systems council have to be industrial engineers. It means that as good managers they should use whatever resources the company has to help them understand the data. They should

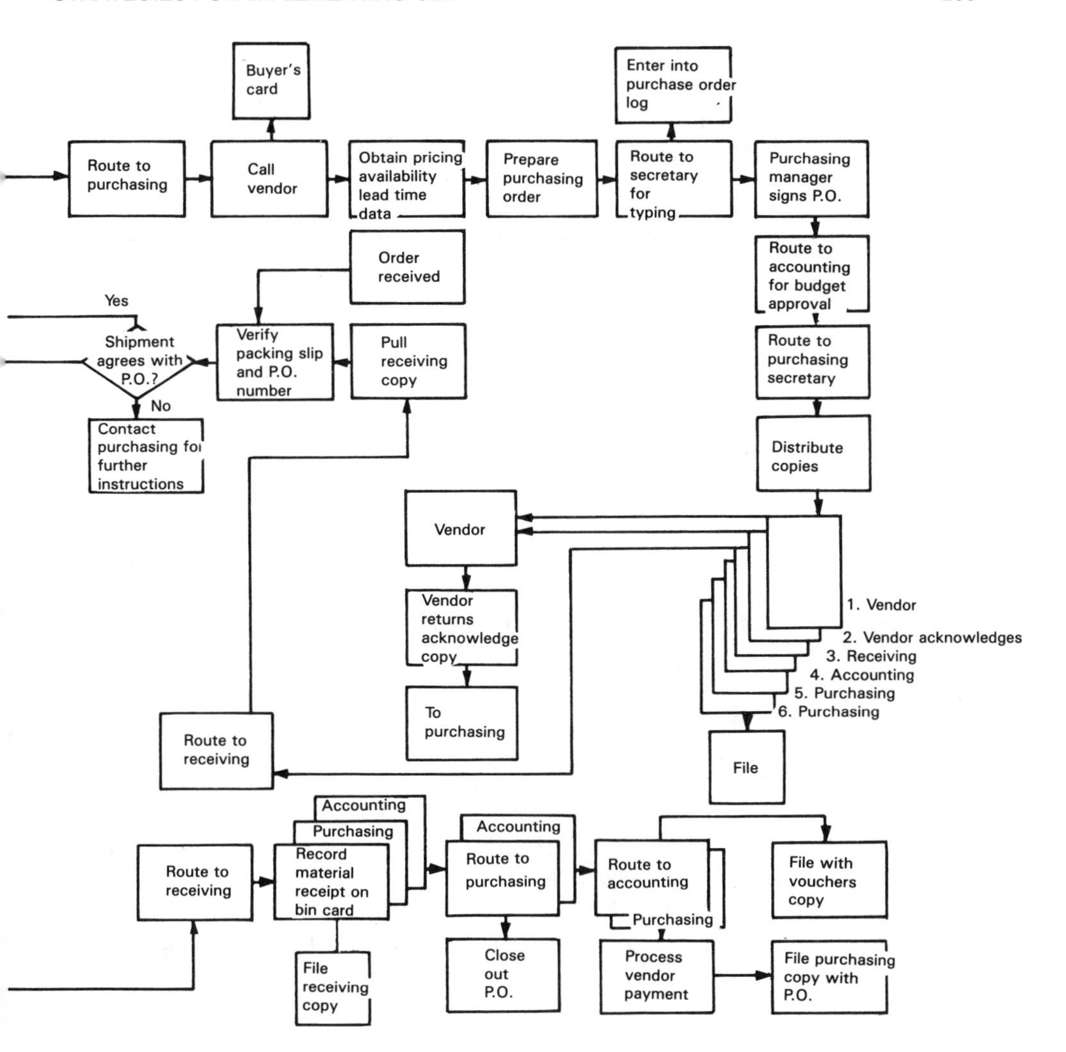

of a brown paper.

even go so far as to hire outside consultants if that is required to obtain the necessary level of expertise. Remember, a project as vital as implementing CIM is not for the timid. If outside resources are needed to properly evaluate the brown papers, then by all means obtain them and use them.

This is probably the most difficult step of the entire strategy. We cannot be mesmerized by the flow captured by the brown paper. This flow has to be critically analyzed for defects, and then the apparent defects have to be assessed to determine whether they are root causes or just symptoms of underlying problems. For documented steps we have to determine if the procedures are cumbersome or too complex. For those of the seven steps of the Manufacturing System that appear to be ad hoc, the problem is even deeper. Why are they ad hoc? Is it a lack of will to codify a procedure? Or is it simply that the procedure was so difficult to follow that it broke down owing to its own ineptness? Whatever the cause, the reason for an ad hoc procedure's being tolerated in use has to be understood.

This critical review has to result in an assessment of what the company does relatively right and what is being done wrong. Notice that there is no place here to say that anything being done is totally correct. This is on purpose. We want to be very critical at this stage and press for improvements in all phases of the present procedure. If we do not, then we will never achieve an optimum solution in the quest for communication excellence. We are tearing the system down to its foundation with respect to the seven steps of the Manufacturing System, and then we keep what is good and enhance it, at the same time building anew with knowledge gained of the benefits of CIM.

Take a moment now to review Figure 2.2. This shows the seven steps of the Manufacturing System laid out with the specific activities that occur under each step. For the brown paper exercise, we attempted to devise how we, as a company, accomplished our variant of the manufacturing system. This is the raw material that we are now analyzing for improvement opportunities. There should be at least one brown paper segment for each segment of the company's specific variant of the manufacturing system. Looking at these segments, our task is to find a way to ensure that all steps of the Manufacturing System are carried out in a fully documented manner that is streamlined and easy to do. With this list of improvements, we will then construct an improved flow for each of these segments, which is an ideal flow for our particular factory.

We have now covered the strategy of carrying out the second task of the process. Let us proceed to the third task and look at some methodology for achieving an ideal manufacturing system that builds on the actual status depicted by the brown paper.

3. *Develop an ideal manufacturing system flow for the particular factory.* Developing an ideal manufacturing system consists of four distinct activities that use as their basis the brown paper and the analysis conducted to understand what it is telling us:

1. Analyze for discontinuities.
2. For discontinuities, find the root causes.
3. Analyze flows for redundancies.
4. Construct an ideal flow.

Let us now define each of these activities and what is involved.

Analyze for discontinuities. This means there needs to be traceability from an information input to an action. The trail should show that an input is received by someone and that individual does something with it. That demonstrates a direct linkage between an input and the subsequent action. Many times we find that there is a discontinuity and we cannot trace an action as a result of the input. Let us look at a simple example.

Suppose a brown paper shows a finished product reaching the shipping area and the next action as placing an order for a truck to ship it on. At first look this appears to be a cause and effect. But is it? Let us list what really happens to point out all the untraced work sequences. This will definitely demonstrate that what we have is not a directly linked cause and effect, but one with many intermediate sequences.

• Finished product reaches shipping area (brown paper, traceable).
• Shipper examines accompanying documents to determine disposition of the product (untraced activity).
• Shipper determines disposition: pack for immediate shipment, temporary hold, or pack for storage (untraced activity).

Now we have three possible routes for information receipt action to trace.

• Shipper denotes immediate shipment required (untraced activity).
• Shipper determines shipment location, orders packing instructions based on location (untraced activity).
• Shipper receives promise date from packers (untraced activity).

I am not going to describe the information flow between the shipper and packer in order to keep this example within reasonable limits. Suffice it to say that this is another example of untraced activity.

• Shipper orders truck to coincide with packer promise (brown paper traceable).
• Shipper denotes temporary hold (untraced activity).
• Shipper places product in temporary hold area (untraced activity).

This will now require another documented flow to show how the product can be released from temporary hold. We will not go into that for simplicity's sake. This is another example of an untraced activity.

• Shipper denotes pack for storage (untraced activity).
• Shipper sends product to packer with instruction to pack for storage (untraced activity).
• Packer packs for storage and returns product to shipper (untraced activity).
• Shipper sends to storage (untraced activity).

This is an example of analyzing for discontinuities. There are certainly enough of them in this simple example, which is typical of too many companies. We fall into the trap of assuming that all members of the company automatically know what to do and then do it efficiently. This is hardly the case and is a source of tremendous productivity improvement opportunity.

For all discontinuities, find the root causes. The shipper example is typical of what will happen when the brown paper is reviewed for the first time. We will find that it is not as thorough as we would hope. For all the untraced activities we now have to determine if these are oversights on the part of those who did the brown paper or if they are really omissions in our system for conducting business. If it is the former, then we amend the brown paper. It is the latter that we have to be concerned with.

With a corrected brown paper, one that truly represents how information flows, or does not flow, we start to analyze why the discontinuities exist. These untraced activities represent the ad hoc method of doing business that most companies fall prey to. For each untraced activity, we ask the following questions.

- Why do we not have instructions on how to do this activity?
- Can all scenarios associated with this activity be described?
- Is the information input for this activity readily available?
- Are the reasons given for the first three questions descriptive of the root cause? Recall that root causes are answers to questions that do not lead to further questions.

This set of questions and possibly some problem-specific additional questions will get us to the reason for the discontinuity. This (these) reason(s) will define an action to be taken to eliminate the discontinuity, the dreaded ad hoc process. Remember, from a managerial control viewpoint, an ad hoc process is like a loose cannon. It is entirely out of control, unpredictable; therefore, the outcome is indeterminate.

Analyze flows for redundancies. At this point we have eliminated ad hoc situations, and at least we can trace the process from start to finish. If we stopped here we would have a manufacturing system that is controllable for each and every activity, but it is far from optimum. Now we have to critically evaluate each work sequence to see if it is really necessary.

What is an unnecessary work sequence? This is a difficult question to answer directly. It will depend on the specific situation. But, in general, a work sequence can be considered unnecessary if it does not have a value-added outcome. Also, a work sequence may be considered unnecessary if when we ask the question, "What will happen if I eliminate this work sequence?" then the answer is, "Nothing will happen if this work sequence is not done." So all work sequences, even though they are traceable (therefore not ad hoc), should be evaluated for

- Value added to the process
- What would be done in an ad hoc manner if the work sequence were not included

The last test implies correctly that if a work sequence is eliminated that is really needed, the organization will invent an ad hoc procedure to get it done.

When this phase is completed, we will have a flow that is as straightforward as possible with very little convolution. Information will come directly to decision makers, and they will have all the information necessary to carry out their tasks. Others may receive information, but it will be strictly for monitoring and information purposes. This secondary loop will be as simple as politically possible.

Construct the ideal flow. With all the information and documentation gained from the first three activities, the systems council can now construct a streamlined flow, called the "ideal flow." This flow chart will show the seven steps of the Manufacturing System being performed in the most advantageous manner possible. It will only have the required information flows, with extraneous information weeded out. The work sequences will be as direct as possible, and all forms of redundancy will be eliminated. Most important, there will be no ad hoc actions because all reasonably expected contingencies have been planned for. As a further protection against ad hoc actions, a procedure will be put in place to ensure that unplanned for contingencies are routed to the appropriate managerial problem solvers and decision makers in a formal manner (examples: material review boards, structured problem analysis routines).

The ideal flow that we have constructed by the four activities represents what we would like to do. The next step is to test it for reasonableness and practicality.

4. *Test the ideal manufacturing system flow against existing specific business constraints.* This streamlined flow we call the ideal manufacturing system is the objective we will strive to implement. It is the one that when fully automated will give us the best CIM system theoretically possible. It will allow us to truly reach the ultimate in achieving communication excellence.

Notice that I use the word objective, not goal. This is not an idle choice. An objective is an asymptotic perfection. In practice it is unachievable. Therefore, the ideal manufacturing system is unachievable. Our task from here on then is to install a CIM system that is as close to the ideal as possible. After we analyze the ideal for practicality of implementation, then we will have an optimum system. This optimum system will be the real goal we seek.

We test the ideal system against pragmatic constraints. Sometimes a company is forced to do things in a less than ideal fashion because it finds itself as part of a larger activity. It may not be operating in an environment of simple cases of vendor/buyer relationships. There may be legal constraints on how the parties relate to each other that forces some redundancy onto the system. This is often true when dealing with government agencies, which have to be concerned with the appearance of being correct. This is beyond the simple concept of just doing the correct thing. When that happens, quite often traceability of processes and materials end up being required beyond all sense of useful practicality. This will probably require more documentation and verification that the documentation has been obtained correctly.

This definitely adds complications to the ideal manufacturing flow; hence, we downgrade to an optimum flow, not quite as good as the theoretical ideal, but the best that can be done and still comply with all the constraints.

Of course, there may be many other causes of backing down from the ideal in addition to contractual ones. For example, a direct flow to a decision maker may not be practical because the decision maker would then be swamped and constitute an unacceptable bottleneck. When that appears, we may have to back down and put in some buffers. Just as the president of a company has a secretary to keep logistical details and protocol from taking up his or her time, an information flow system may need a similar buffer. Whatever the reason, we need to thoroughly test ideal flows for practicality and determine what is accomplishable. Knowing what is possible is a necessary precursor for developing an optimum manufacturing system.

Finally, the practicality of achieving an ideal flow has to be tested against the state-of-the-art of software and computer hardware. We certainly can conceive of flows that give us instantaneous information in any format we wish, but can it be done with the currently available technology? This is a true pragmatic constraint. Ideal systems should be conceived that stretch the current capabilities of computers and software.

It is this testing and probing that defines what can be done now and what may be possible in the future. Without this stretching, progress will not be made. All we have to do is think back perhaps to the mid-1970s when relational databases were not available. That led to constraints requiring as much preconceiving of possible outcomes as possible to build network database algorithms to handle them if they ever occurred. Obviously, we were not capable of foreseeing every possible scenario, so our manufacturing systems were not as thorough and effective as today. Today we have relational databases that make programming simpler and responses to unexpected events more effective than ever before. But, to have designed ideal systems requiring the capability we have today was the proper thing to do. Sure, we had to back down to a less effective optimal system that was available. But to have done otherwise would not have spurred development of what we know is the state of the art today. It is necessary to design the ideal, test it against practical reality, and only then reluctantly back down to what is currently available. This ensures that the optimum system we eventually employ is the best possible and that development work is focused on what is really desired.

5. *Based on constraints, develop an optimized Manufacturing System flow.* Once we have tested the ideal Manufacturing System against the real world constraints we are faced with, we redesign it to be compatible with those constraints. We strive to be true to the concepts of the ideal system, and by no means should we forego basic needs because they will be tough to accomplish. However, we must be cognizant of what is currently possible vis-à-vis technology. Logically, it will do no one any good to hold out for future developments if those developments are beyond reasonable expectations. This means that the optimum system we are downgrading to is the best available at this time and so we will settle for it. The only other option is to abandon CIM altogether and just wait for technology to catch up with our desires. This is hardly a preferred option. Even though we must compromise, we must never give up on the ideal. We must make sure that the optimum system we settle for is not dead end technology and has room for implementation of new technology when it becomes available.

This business of constructing an optimum manufacturing system flow requires a lot of give and take. It will probably require many iterations before the systems council is convinced that it has the best system for their company. How do we go about formulating an optimum manufacturing system flow? Much as we did for the previous two flows. The ideal flow had as its predecessor the brown paper. In a similar manner, the optimum flow has the ideal flow as its precursor. We again question every segment of how the seven steps of the Manufacturing System are accomplished, only this time we are looking at how practical those ideals are. For accomplishing each step of the Manufacturing System we ask

- Is the technique compatible with the current common database concepts?
- Where will the data come from?
- Is the information source capable of supplying the data?
- Is the amount of data within the realms of computer manipulation? Of memory and disk storage size bounds?
- Is the simultaneous usage requirement within the bounds of current technology, or is the proposed need likely to mean very slow response times?

• Is the requirement for data display and collection within the realm of technical feasibility?

• Is the cost reasonable with respect to the company's resources?

Note that none of these questions refer to the basic questions of streamlining flow or eliminating redundancy or changing ad hoc conditions to preplanned situations. Here we are looking at the practicality of doing what we would like to do. All the questions are primarily computer systems feasibility questions. This is because the computer hardware and software capability will most likely be the weak link in doing what is desired. Also, the last question has to be answered honestly. There is no sense in contemplating a system that if purchased would bankrupt the company. Chapter 13 covers economic justification, and those topics must become part of the systems council's deliberations.

Figure 14.2 illustrates the path the systems council will take in evaluating the ideal to formulate the optimum.

We can see that the primary evaluation requirements have to do with size requirements of the database and the ability of the computer hardware and software to accommodate it. This means the systems council has to really accomplish the task of sizing the database. We do this by literally counting the inputs and outputs to the database that the various components of the company will require.

This is a first shock of many more to come to the employees of the company. We are now asking them to think in terms of database inputs and outputs. Previous to this we have philosophized with them about change and how computers will help them do their job. The requirement to structure their needs in terms of input and output will be a social shock to them. This will be the first indication of direct changes to come in the way they carry out their assignments. Figure 14.3 illustrates what I mean. All information is input to and received from the storage receptacle, the common database.

The specifics of what the flow mechanism will be is not important yet. That will be decided after the systems council determines what the volume and flow rate has to be. (See chapter 3 for discussion of the merits of the various internal and external software systems architecture choices available.) Determining the flow rates and volumes of information tends to scare people, and it should not. We simply count the number of typical information transactions over a representative period and then, by a direct correlation of computer memory bits to letters, numbers, symbols, and spaces, we add them up and see how large the computer memory and storage disks have to be. We then compare this with available commercial offerings, and we see if it is a match or a problem. If it is a problem, then we have to downgrade the ideal requirements to what is practical. That is the core of the task that the system council has to perform to judge whether or not the ideal is feasible. We also have to take care of the contractual and political requirements as discussed previously.

What is the best way to handle this counting of "message units" task? I have found that the most straightforward method is the best way to accomplish it: Simply provide each component of the company with an input and output tally sheet. Label one column "information needed from the database," and label the second column "information

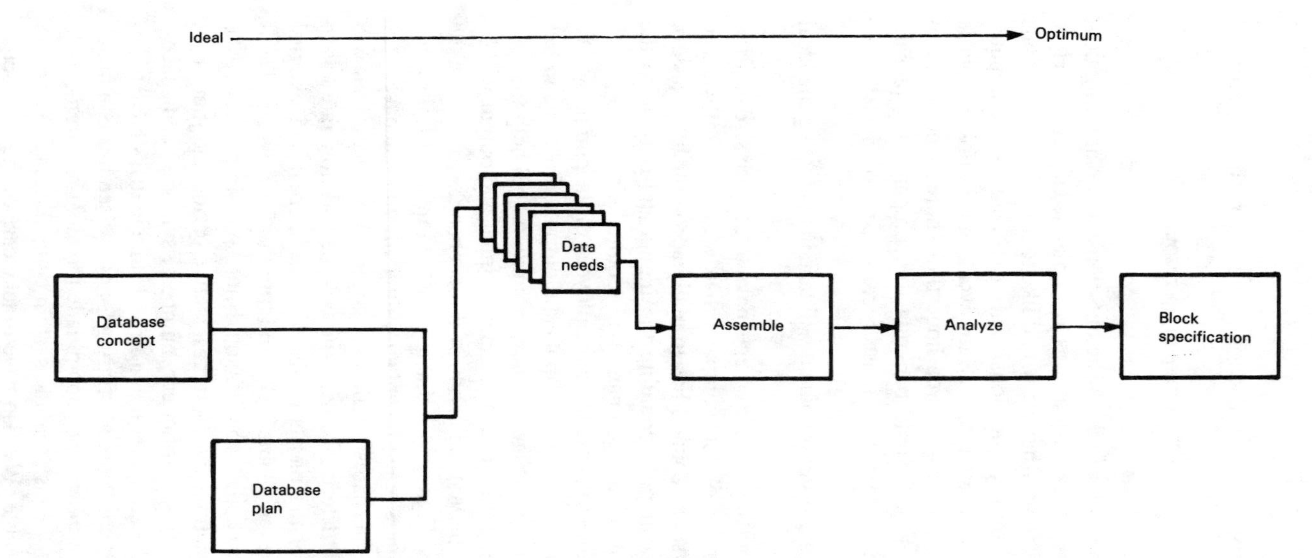

Figure 14.2 Going from the ideal to the optimum Manufacturing System flow. (From B. W. Sallard Consultants, Trumbull, CT.)

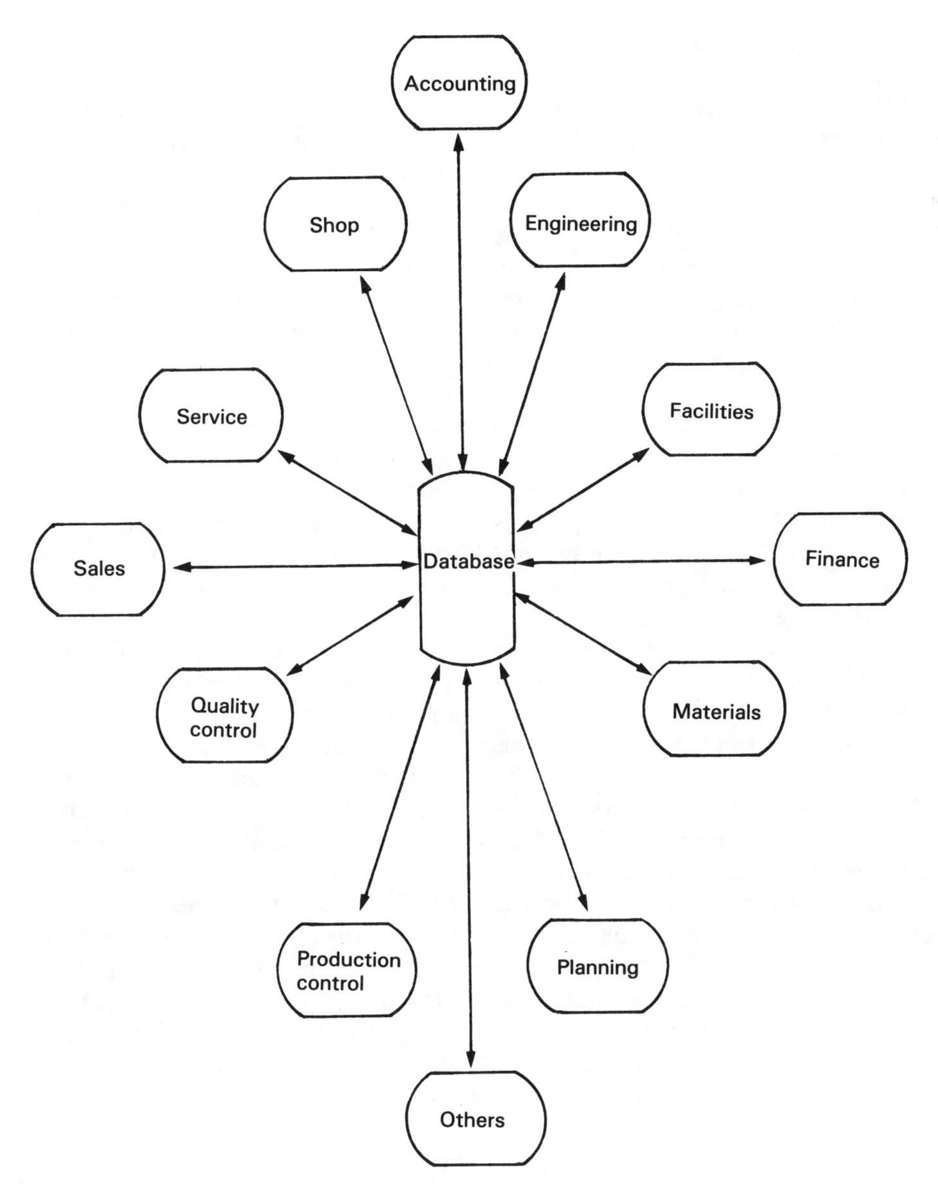

Figure 14.3 The database concept of information flow.

supplied to the database." Ask the prospective users to complete this list to the best of their ability, and collect and collate. I would also recommend that the individuals supplying the data be given a briefing concerning the reason for doing it. Perhaps also some suggestions about how to go about the task are in order. We ought to be suggesting that daily logs be kept, or files reviewed, and probably the most pragmatic thing to do is to review the ideal flow with the participants so they know what the systems council contends ought to be happening in their component. This type of communication will result in sufficient data being collected.

The idea of reviewing the ideal flow with cognizant company components has more uses that just gathering data. It is a way of nurturing participation by a wide range of diverse personnel. This participation will greatly benefit the implementation phases of CIM. If we have a fully informed work force and if we incorporate as many ideas as possible from this work force, then the CIM philosophy we are trying to install becomes "our" plan instead of "their" plan. The chances of being successful with "our" plan are much much greater than with "their" plan. "Our" plan has built-in ownership and therefore interest in its success is guaranteed. "Their" plan may grievously suffer from NIH (not invented here) and could be subversively opposed.

When the task of evaluating the ideal for pragmatic applicability is completed, the systems council gains the capability of structuring an integrated computer-based manufacturing system. With the optimized manufacturing system flow, the systems council understands the best way its company can accomplish the seven steps of the Manufacturing System. The next phase is to take this manual or semi-computerized system and transform it into a CIM system. What has to be done is now, for the most part, known, and the danger of computerizing an inefficient or non-workable system has been sufficiently minimized.

6. *Select the order of implementation of subsets of the optimized manufacturing system flow based on company needs and financial and technical resources constraints.* The effort up to this point constitutes a major portion of the creative work required to implement CIM. The rest of the work will be matching these specifications, as shown on the optimum flow, to existing or to be developed software and hardware and then implement within the company (See Figure 14.4). The sixth task is heavily factored by available resources. The systems council now has an overall design concept that appears to match the realities of the company's situation. How fast the concept can be translated into a firm design, then implemented, will depend on the financial and technical people resources. This means the systems council has to set priorities.

How to set priorities depends on the company's philosophy and needs. Something had to trigger the desire to enter the CIM arena. Usually it is a desire to solve a problem that defies solution by any other method. Typically, companies just do not say, "Hey, let's do CIM this year." Companies are desperately trying to solve a problem, and they sort of drift into the CIM universe. They really do not know anything about CIM at all as they start to get involved. Instead, they have heard or read about good things that have come to companies that are using computers to solve problems. So, this by-chance introduction to computers leads to better understanding of the communication excellence approach, which in turn leads to the discovery of CIM and all it can do for the competitive position of their company. I have found that companies become involved with CIM for two distinct reasons: (1) to solve serious quality problems and (2) to solve serious productivity problems. Each will lead to a different approach to setting priorities as to what should be implemented first.

The quality problem approach typically arises because the firm is having trouble building its products to the desired specifications. These could be yield problems, maintaining tolerances, and perhaps just being able to build their product as well as

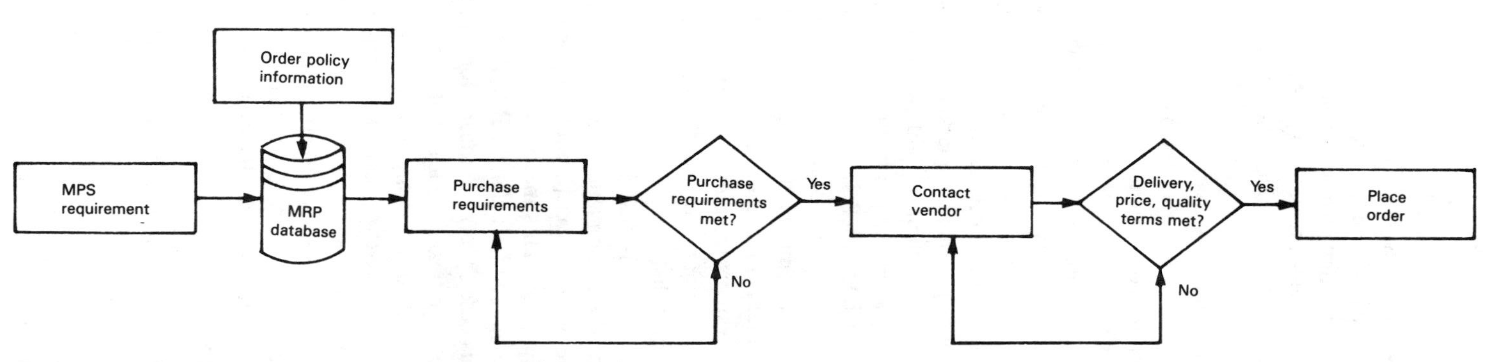

Figure 14.4 Example of an optimum Manufacturing System flow.

their competitors. The solutions first thought of include machines and processes that can be held to tighter specifications. This, in many people's minds, is computer control. So, the first priority from this viewpoint tends to be the machine/process control component of the CAD/CAM triad. They emphasize controlling the process through linked N/C machines, the FMS, and robotics. Soon, they realize that the equipment by itself is not a sufficient solution. So we start to see elements of the production and measurements control component being introduced. Usually statistical process/quality control is first, then the natural evolution to MRP II, and finally the design and planning control components are seen to be beneficial and a necessary part of the system. So, companies with perceived quality problems would tend to set priorities in this manner.

Companies with productivity problems look at priorities in an altogether different manner. They are obsessed with the problem of not being able to make enough of their product. What they make is usually excellent. Their problem is one of not being able to hold their costs down because costs cannot be spread over enough volume, or they cannot get or hold market share because they cannot meet their customers' quantity needs. So, their introduction to CIM most often is by MRP II or JIT implementation. This means the production and measurements control component is their first priority. Once into MRP II, they find that bottom-up CAPP is just as necessary and so is integration of design with factory planning. This results in the design and planning control component being drawn into the evaluation of priority. Finally, productivity problem companies reach an inevitable conclusion that if they had N/C machines and derivatives working with MRP II, then they would make even more quantum jumps in productivity. So, we see machine/process control being the last component on the priority list.

What is described here are the two schools of thought for setting priorities based on where companies saw a need for CIM. Probably, most companies will favor one or the other approach, but most companies will also have needs that are satisfied by both scenarios. This means there are no hard and fast rules as to how to set priorities. I recommend a balance. Recall that throughout this text I have stressed synergism. In this case synergism means for optimum results we cannot afford to let one component get significantly out of balance with the others.

I believe timing is more important than the particulars of what comes first. On the basis of my experience I can offer this piece of advice: The longer the implementation time frame is, the less likely the company will fully implement the program. Usually two to two-and-one-half years to implement is about the maximum time a company should take. Any longer than that and the corporate attention span starts to drift, which in real terms means the money will start to dry up. So, schedules ought to be established to do the entire program within a nominal two-year time frame.

Another factor in setting priorities is the economic one. Obviously, the projects selected to be done have to pass the company's justification procedure. Since this is covered in depth in chapter 13, I will not go into it here. However, the systems council must keep in mind that CIM projects are harder to justify because precise justifications of savings are harder to do than for many other capital investments. Much of CIM is based on improved communications capabilities, which yield significant savings. It is difficult to pin down precise cause and effect. This means a lot of the savings of individual projects will be considered soft. There is definitely a challenge to the systems council to extol the benefits of CIM. By doing a good job in the selling aspect, the justification problems become easier to handle.

Finally, the priorities and pace of application will depend on the technical skills the company can muster. To implement CIM requires a core of engineers who understand the technology and are conversant in the various software package applications. It is necessary to have on board sufficient numbers of these people before, during, and after implementation. The systems council will establish the priorities, but the technical staff will do the implementing. Part of the planning the systems council will have to do is plan for the types and numbers of engineers required to implement the program. On the basis of financial resources available, the systems council will have to balance a desire to implement as quickly as possible with the realities of the economic/technical constraints.

We see, then, that priorities depend on company perceived needs, financial constraints, and technical constraints. It is the systems council's responsibility to balance these three factors to come up with a plan that implements the CIM program as quickly as possible.

7. *Use the prototype process with vendors to modify their existing software offerings to comply with the optimized manufacturing system flow and create an acceptable integrated computer-based manufacturing system.* Up until now we have only constructed an information flow that we desire to use as the basis of a series of integrated software programs. Now we have to devise a way to have software written to meet the company's needs. That process is called the prototype process. Let us look at that process.

Once priorities are established and approval to commence given, the order of implementation from the optimized manufacturing system flow to the integrated computer-based manufacturing system is set. Now the process of implementation begins.

The start of this process is the determination of whether commercially available software and hardware systems suit the needs of the company. Ninety-eight times out of a hundred, commercially available software is not suitable for use without moderate to serious modifications. Hardware decisions tend to follow the choice of software for new computer users and tend to be a constraint for old computer users. For those companies already owning computer hardware, there will be a strong economic desire to use it for future systems, if it is not hopelessly obsolete. For those companies not encumbered with existing equipment, their path is to find the best software possible and match it with hardware that it efficiently runs on. In either case, the job is to find the best software within whatever the constraints may be. For this reason, the implementation process always starts with the selection of software.

Using the optimized manufacturing flow and its associated data input/output inventories (see Table 14.2 for an example of what kind of information may be found on a data input/output listing), the systems council will arrange to review various vendor product offerings for their ability to achieve all the system flow requirements.

The best way to do this is through the prototype process. Most vendors detest this method of software selection, therefore it must be pretty good. Vendors dislike it because it forces them to evaluate their software packages against very specific require-

Table 14.2 A Data Flow Requirements Submittal
CIM Project data output/input listing

	Output from Database	Input to Database
1	Absentee reports	Actual delivery date
2	Account number	Actual production information
3	Accounts payable	Actual retail sales by month
4	Accounts receivable	Actual retail sales by quarter
5	Actual operating time	Actual retail sales for the year
6	Actual production history	Actual w/s load: units/time period
7	Arrival date	Attendance control reports
8	Attendance history	Budget information
9	Attendance records	Check register data
10	Bill of material	Commissionaire's address 1
11	Bill of material: multi level	Commissionaire's address 2
12	Bill of material: single level	Commissionaire's %
13	Bill of material: summarized bill	Construction order number
14	Bills of material	Customer service files
15	Budget estimates	Cust. account number
16	Budgets and actuals	Cust. account type
17	Budgets and actuals expenses	Cust. city
18	Budgets and actuals headcount	Cust. company name
19	Budgets and actuals supporting data	Cust. zip code
20	Burden dollars (act)	Cust. name/title, if applic.
21	Cartage cost	Cust. sales order number
22	Construction order list	Cust. state
23	Contract dollars	Cust. street address
24	Control: critical resource capacity	Cust. telephone number
25	CPR's in process	Daily sales by commissionaires
26	Critical resource requirements	Daily sales by sales reps
27	Current calendar	Date of sale
28	Customer address	Date report printed
29	Customer city	Department description
30	Customer file	Department number
31	Customer last name	Drawing revision
32	Customer middle initial	Drawing size
33	Customer name	Effective date of eng change
34	Customer orders	Engineering change request number
35	Customer state	Expected delivery date
36	Customer telephone number	Fringe benefit accruals
37	Customer title (Mr. or Mrs.)	General ledger journal entries
38	Customer zipcode	How much material to order
39	Daily actual production	Labor distribution
40	Daily time reporting	Lead times for parts
41	Daily time reporting by department	Master schedule
42	Daily time reporting by operation	Monthly inventory needs for retail
43	Database input/output listing	Monthly status of expense accounts
44	Date completed	Monthly status of promotionals
45	Date received	Notification of change number
46	Date shipped	Operation description
47	Demand: actual orders	Operation number
48	Demand: forecast	Operation sequence number
49	Department efficiency	Part classification

ments. It makes the vendors work for the sale. It forces them to really understand the customer's needs before they can respond and quote on the job. This is in contrast to the usual sales technique, that of convincing the buyer to purchase what the vendor has and not what the customer wants.

The opposite of the prototype process guarantees that the buyer will have to make major modifications in the way the company does business in order for the software to work. The results will be an unhappy customer and a discredited vendor. I have seen this happen over and over again, and the reason is simple. A software company cannot design software that is both specific and generic at the same time. Since the different specific uses of software can be vast in number, while the generic philosophy of a program can be universally acceptable, the designers have no choice but to offer generic process software for sale. Generic software may represent a significant percentage of usefulness, but it never fits the bill entirely. It requires modification before it can be used at a specific company.

Think of generic software as a suit of clothes from a fine store. It is rare indeed for a person to go into the shop, find a suit, and have it useful without some alteration. We also know that not all clothing off the rack can be altered for every customer. A person requiring a 46 long would have a low probability of success with altering a 42 portly to fit. The same is true of software. We can start with a near match generic offering and with altering make it work without changing the way the company chooses to do business. But if the software is not close enough to the needs spelled out by the systems council, alterations alone will not be suitable. This should never be accepted, even at the urging of respectable vendors to change the desires of the systems council to allow the software to fit. An altered 42 portly will never be satisfactory to a 46 long.

The prototype process is the best way of making sure that only near matches of generic software are considered. The process is in two phases. The first allows all vendors to state how their software already meets, or can be modified to meet, the requirements defined by the optimum flow. The second phase culls out the obvious "no matches" and focuses the systems council on the two or three vendors with potential for satisfying the needs. It is working with the multiple vendors at the same time that is the heart of the prototype process. This is a process of mutual learning between the systems council and the vendors. It is an iterative process that greatly reduces the time period required to implement software and, quite frankly, is probably the only way a firm can introduce multiple modules required for CIM in two years or less. The systems council undertakes to teach the vendors its needs vis-à-vis information flow. At the same time the vendor is instructing the systems council on how its software works, in particular, how the package satisfies the needs of the firm specifically and what portions will need to be modified. This iterative process guarantees that the modified software eventually purchased really does come as close as possible to meeting the original flow specifications. The firm will usually only miss out on what is still beyond the state of the art of software design. This means everything that is possible to do will be done. The prototype process generates a much higher level of satisfaction than going with the normal method of accepting what is available and modifying the company's methodology to comply.

Working with two or more vendors at the same time, in essentially a design run-off competition, can be tricky. It is difficult for the vendor because he or she may put in a significant effort and not get the job. It is equally difficult for the buyer because a major commitment of company resources has to be expended in duplicate and perhaps triplicate (and not simultaneously) to give each vendor equal access. The systems council has to make sure they scrupulously observe the standards of professional ethics. This means they must be on guard to not divulge proprietary information given by one vendor to another. They must also not allow themselves to fall into the vendor baiting trap, where they imply the other vendor is willing to do something in order to get this vendor to offer a concession. Fair play is essential. If the firm finds it cannot do this, then it should very quickly eliminate all but one vendor and work the iterative process only with the selectee. This is not optimum, but it is a wise choice if ethics violations could become a problem.

The prototype process is an interesting and vital methodology. Let us go through a descriptive example to illustrate how it works. We will use Table 14.3 as a guide for describing the technique.

Let us set the stage. Suppose we are company X and after much deliberation have set up a systems council to get our company into the CIM age. We have progressed nicely up to this point and are now ready to express our information needs to the suppliers of computer software. We are very wary of this cunning breed and want to make sure we get what we need. We have heard horror stories of companies that relied solely on the vendors to tell them what was required and subsequently failed to meet their goals. We cannot let this happen to us. We are very willing to pay a reasonable price for the software, and our company management is supportive all the way for implementing the right system. This means we cannot disappoint our management by implementing anything but a CIM system that meets our specific requirements. We know that this move to computer usage is a big change for our company, and we cannot tolerate the additional shock of changing the way we like to do business in order to get software implemented. We want software vendors to work with us to make the change over to a computerized information system as painless as possible. Therefore, we've chosen the prototype process as a way of achieving our goals.

Table 14.3 The Prototype Process

1. The systems council presents the optimum Manufacturing System flow to potential vendors.
2. Receive preliminary proposals from vendors for meeting requirements (no cost estimates included).
3. Make technical evaluations of preliminary proposals.
4. Select two or three vendors to work with toward receiving full proposals.
5. Commence iterative education activity, one on one, between systems council and vendors.
   a. Systems council presents information flow needs to accomplish each of the seven steps of the Manufacturing System and linkages between them.
   b. Vendors relate what their software can do to accomplish the specific intent of each of the seven steps. What can be done with no modification, some modification, and what cannot be done.
6. After the education activity is completed, the vendors prepare technical and economic proposals.
7. The systems council makes technical and economic evaluations of the proposals.
8. A vendor is selected.
9. The systems councils and the selected vendor agree to an implementation schedule.

Step 1. Presentation to vendors. The first step of the prototype process is to present our optimum manufacturing system flow to invited vendors. We have invited ten likely vendors, a combination of hardware producers (virtually every major computer producer also has extensive software capabilities to support their offerings) and strictly software supplier companies. To make our work as easy as possible, we have decided to make one presentation to all the vendors and have agreed to answer their questions either in assembly or later in private.

Since our systems council is made up of representatives of all phases of our business, we have divided the presentation responsibilities among our functional representatives. Each of the seven steps of the Manufacturing System will be handled by "our" cognizant expert. Also, to make sure there is no doubt on the part of the vendors as to what we mean by linking the steps of the Manufacturing System together, we have decided to show our entire optimum flow on one chart. We are doing this by laying it out on one very long brown paper (it is not unusual for the optimum flow to be fifty to one hundred feet long). To do this, we have hired the ballroom of the local hotel for the day and have tacked the brown paper to the wall.

Our method of presentation will be to have the vendors tour the brown paper with us around the room. Each presenter will stand in front of his or her portion and explain it in as much detail as we think can be absorbed by the vendor. At the end of each presentation we will answer questions and hand out copies of the flow with the appropriate data input/output requirements. By the time we get through the section of the optimum flow representing the last step of the Manufacturing System, each vendor will have a complete set of flow charts and data input/output requirements. We will also give them any general requirements we deem necessary. We then ask the vendors to give us a preliminary proposal, within three weeks, which spells out how they would go about solving our problem. The tight time frame is reasonable. This will weed out any vendors who do not already possess the capability of satisfying our needs. We do not want to be the first users of brand new software.

Step 2. Receive preliminary proposals. Having successfully made the presentation of the optimum flow, we now make ourselves available to the vendors to answer their question. They will have two types of questions: technical ones and how the chosen vendors will be selected.

The technical questions we will answer as honestly and completely as we can. The question on vendor selection has to be handled differently. We certainly do not want to divulge the specifics of how we intend to rate the proposals, because we want the vendors to be as honest as they can be showing us how they will meet our needs. We do not want to fall prey to slick salespeople who will appear to tell us what we want to hear and yet not deliver the specifics they allude to. If we tell vendors specifically how we intend to rate proposals, this is exactly what will happen. It is not in our best interest to inform the vendors specifically how we intend to rate their proposals. To ensure the best response possible, we answer the selection question by stating that their proposal will be judged on how well we think their proposals will meet our needs, as described in our presentation.

Step 3. Technical evaluations of preliminary proposals. We have chosen to use a variation of the "musts and wants" technique for evaluating proposals. Previously, the systems council has drafted a set of "musts," which spell out the absolute items the proposed solution has to be able to do. If any of these are missing, it is automatic disqualification for that vendor. This is like the hurdle rate for economic justification evaluations. Along with the "musts," the systems council has also drafted a set of "wants." These are nice things to have but not absolutely vital to success or failure of the software. We also gave each "want" a rating, a 10 for being the most desirable down to a 1 for being the least. When we rate the individual proposals, we will add the score of all the "wants" the vendor has included in its proposal.

The winners of the bidding will be the vendors with the two or three highest "wants" scores. We have already eliminated the vendors who cannot satisfy the "musts."

Our systems council decided to break the "musts" and "wants" into two categories. The first category was the technical category. This activity was led by our member from manufacturing engineering and our advisor from management information systems. They used the resources from their parent organizations to develop the technical details for this portion of the "musts" and "wants." Their list contained items such as terminal response times, computer degradation rates with respect to the number of users accessing the memories at the same time, the number of lines of information that could be presented to operators at any time, and software command logic sequences. They had several pages of "wants," but the systems council restricted them to a maximum of only ten "musts." We did this to make sure the "musts" were truly necessary from an objective point of view and were not simply emotional desires.

The second category dealt with managerial and general issues. We again limited the number of "musts," for the same reason. Some of the "musts" company X needed were an MRP II package compatible with the JIT philosophy, a bottom-up CAPP system, and a SQC capability compatible with the CAPP system and usable with the proposed data collection system. Like the technical portion, we had several pages of "wants." The few 10-rated "wants" were relational database for ad hoc query capability, real-time update of production control activities, a maintenance database linked to the quality database and sharing the same problem analysis programs, and an electronic messaging capability tied into all databases based on the MAP/TOP protocols. Our company believes very strongly in having detailed control of our operations and in all managers being able to query current activities.

Step 4. Select two or three vendors to work with. After three weeks our systems council is meeting to review the preliminary proposals. We have received five replies from the invited vendors, plus a visit from a sixth by their senior sales executive, who first made a case for us to give his company the job because he knew they were the best for the job, and our doing so now would save us a considerable amount of time. Besides, he knew his company could do the job at less cost than the others because his software was exactly what we needed. His company could guarantee that they could teach us to use their software and we could become very profitable. Only substandard personnel would fail to learn their wonderful and proven system. We declined this offer. The senior sales executive then proceeded to tell us that the prototype process was wasteful, and responsible concerns, such as his, would not lower themselves to participate in such a farce. After all, his firm and others like them knew the business, and clients should trust their good judgment to provide just the right software for company X. We thanked the sales executive for taking time out of his busy day to come visit us and bid him good day.

We had heard about vendor reluctance to compete by the prototype process and about salespeople willing to sell anything regardless of whether or not it was applicable. We now had an example of that. The four other companies that failed to bid fell into two categories. Two said they did not think their software was close enough to our needs and therefore declined to quote. The other two just did not bother to respond at all. Perhaps they too did not believe in the prototype process.

Our systems council diligently reviewed and scored all five responses. One did not meet the "musts" test and was immediately eliminated. Of the other four, one was significantly below the rest and was eliminated without much discussion. The other three had fairly close scores, and we could have decided to proceed with all three through the next steps. But we opted to work only with the top two, because we did not think we could afford the time to deal with more than two vendors and still do an adequate job educating and being educated. We also thought that the proposals were close enough that we would not lose much by working with only two vendors.

We thanked the three losers for submitting their proposals and regretted to inform them that we had chosen others to be finalists. We did not intend to tell any of the losers why they had lost. Only one approached us to ask why, so we gave the salesperson our frank and honest reasons. We thought we might want to work with this vendor someday and to do this demonstrated that we were fair with them and respected their concern so they could do better next time. We informed vendors A and B that they were chosen and set about finalizing the logistics of starting the next phase.

Step 5. Iterative education activity with the finalists. We now knew that one of these two vendors would become a very important and long-term partner in our company's future success. Therefore, we were determined to do a good job transferring knowledge of our needs to these vendors. We hoped that they would be reciprocal. Before we could get started, we required each vendor to sign a proprietory information protection agreement. They each did and in turn asked that we do so also for protection of their company's confidential data. We were pleased to do so.

Our plan was to go over each step of the Manufacturing System delineated in our optimum flow and explain every aspect of it. We would have loved to give them exams to make sure they really understood what we meant, but that would not have been professional.

After each of our presenters finished preparations for his or her part of the tutorial (the same systems council members who made the initial presentation), and presented its content summary to the rest of us, it became apparent that we were assuming something that was not necessarily true. We had assumed that these vendor sales and technical representatives understood the ramifications of running a manufacturing activity. This is not so. These people, very well educated in their software product, are not manufacturers. Even our own marketing people were much more attuned to what manufacturing companies do than the vendors' personnel. They did not come from a manufacturing culture. We asked colleagues in technical societies and universities, and they confirmed our suspicions.

While computer and software companies cater to manufacturing clients, their main business is with financial, distribution, and service companies. Unfortunately, selling manufacturing-oriented business software is still relatively new and not a major factor in their businesses. They may have some personnel who have manufacturing backgrounds, but even these people are primarily from what we call support functions (e.g., management information systems functions, etc.).

Therefore, we realized that we not only had to teach these vendor personnel what we needed for an integrated computer-based manufacturing system, but before that could be successful, we had to teach them the precepts of manufacturing. We would have to delve into the theory of the seven steps of the Manufacturing System, the two knows, and aspects of queuing theory so they would understand the vital intricacies of production scheduling. More important, we would then have to bridge the gap between the seven steps and the CAD/CAM triad. Even more important, we would have to explain the concept of communication excellence.

This is preposterous, some of our colleagues said. How can a computer vendor not know about communication excellence, the CAD/CAM triad, and the seven steps? These, after all, we believe to be the basis of CIM. But, why should they know about these things? They should not because they only write software, they are not experts in its uses or why a client wanted it that way. They simply do a better job writing program code than we do. We do a better job understanding and managing the intricacies of a factory. There is no more reason to suspect that they know about CIM than that a flight attendant can pilot a 747 airliner. The attendant spends considerable time in flight but does not profess to know much about how he or she got up there. A software salesperson and his or her technical support spend a lot of time supplying CIM-related software, but that does not mean they need to understand why it has to do what it does, other than the obvious: that the customers seem to want it that way. This, we came to believe, was the reason the off-the-shelf software rarely meet the true needs of the customer. The client did not know how to ask for what he or she needed because there was a lack of knowledge of computers and software. The vendor, being equally ignorant on the opposite side, did not really know what to offer. We are determined to not let this happen to us. We want to educate the vendors about our business, and we want to learn about what their software can do and how it works. If we can do this, we will have a satisfactory proposal from both vendors to evaluate.

Just as we intend to educate the vendors on how manufacturing really functions, including the basic theory of manufacturing, we want these vendors to educate us about the theories of software structure and development. We want to understand what the vendor's software is capable of doing, and what it really means to make modifications. This way we can make judgments as to what some of our cherished whims may really cost in terms of dollars and performance. We want to be able to make our final vendor choice based on knowledge, not emotion.

We want the education sessions to be a positive experience for all parties. We intend to do this by having the sessions last four hours, with a break between. Each side, client and vendor, will make a two-hour presentation in accordance with a schedule worked out in advance. We will take turns going first to balance the giving and getting. Also, to allow vendors and ourselves the luxury of not having to record notes profusely, we have arranged to have the sessions videotaped. This will help us later if we need some reference to what the vendors are stating when they present their proposals. One other thing we will do is to make the inputs to the vendors as fair as possible. We will strive to cover the same material on the same or adjacent days with each finalist and alternate who is exposed to it first. We expect to have at least four sessions with each vendor, so we will be quite busy for a few weeks.

Vendor A invited us to view a similar application installed in another company. While we would definitely like to see this, we felt we ethically could not unless vendor B was willing to do the same thing. We broached the subject with vendor B, and they too offered to host us at an application site. So, we will be supplementing our education with two field trips.

Step 6. After education sessions, vendors prepare proposals. We have finally completed our education seminars and now have given the vendors one month to prepare their final proposals. They have an excellent understanding of what we require. We have no doubt that they will present proposals that are specifically tailored to our exact needs. This will make the selection process difficult for us, but, even if we make a mistake, the chosen solution will fit our needs better than anything we could have conceivably received with a non-prototype process selection.

We know now, having experienced the prototype process, that we have succeeded in achieving two sets of specifications that are unique for addressing our CIM needs. We could not have done the job without the vendors because we certainly did not possess the in-depth knowledge of software and hardware applications pitfalls and advantages. Neither could we have received as good a specification from consultants, because they would not have had a complete knowledge of our business in order to do such a detailed and thorough job. At least it could not have happened as fast. We are now three months into the vendor selection process, and we have a specification that normally would have taken more than a year to produce. Even then, it probably would not be as complete as these specifications are. Adding the four months it took to get to the optimum flow stage, we are still ahead of the point we would be with any other approach.

The systems council is satisfied that we have minimized future implementation problems by having a much more detailed specification to use with the vendors. In fact, we are sure that the vendors feel the same way. Each vendor has their own specification developed with us to bid on. They should have fewer problems implementing because we understand each other far better than in the usual vendor-client relationship. There is another benefit that will accrue to us. Since the vendors know with much greater accuracy what we want, and how we will work with the supplier, they can sharpen their pencils and not have to build in such large fee contingency values for unknowns. Of course, we understand that they are going to charge us somehow within their proposal for the time spent on the prototype process exercise. We know it will be far less than the savings for less contingency funding. Besides, there are two vendors in the competition, and they would each be reluctant to lose the job by being too greedy in recouping their sales costs.

Step 7. Evaluate proposals. A month later, on the due date, we receive two very impressive document sets from our vendors. The moment of truth has come, and now we will learn if the prototype process has really worked or if the contestants have fallen back to their previous ways of telling us what we need, ignoring the facts of the situation. To our relief, that is not the situation. Now the systems council has to get down to business and make the evaluations.

We pull out our old "musts" and "wants" folder to do the evaluation, only this time it is slightly different. As we listened to the competing vendors, we came to the conclusion that some of our "wants" were rated incorrectly. After the education sessions, while we were waiting for the proposals to be completed, we viewed the videotapes for more insight on the merits of the "wants." We played devil's advocate with each other and honed the "wants" set to greater practicality, based on our newly acquired education. We even added a few new "wants," something we could not possibly have done if we were stuck with a non-prototype situation. We very carefully do the evaluations and come to a conclusion. But we do not announce it!

To be sure we have picked the right vendor, we decide to present both vendors' proposals to our fellow managers and senior managers in a series of small group meetings. We do this because we want to get feedback and make sure our decision is not an isolated one. After all, we have been in close contact with both sets of vendor personnel for quite a while now, maybe even closer with them than with our own people. We make our presentations and take careful notes of our colleagues' comments.

Step 8. A vendor is selected. With the feedback from our proposal roadshow, and our previous evaluation work, we caucus. Is our initial decision still valid? Have the reactions to the two proposals caused us to change our collective minds? After careful consideration we reach our final decision. We announce our decision to our senior management, with our reasons. We ask for their concurrence, and, after receiving it, we are ready to tell the competing vendors.

We handle this the same way we did in selecting the finalists, professionally and with tact. Our leader first calls the loser to inform that firm we have decided to go with the other vendor. They ask why they lost, and we tell them as diplomatically as possible. We even agree to go to lunch with their team leaders within the next few days to explain in detail why they were not selected.

Dealing with the loser is not easy to do. We have built up a rapport with these people over the past few months, and it is difficult to let them down gently. We believe we owe them this, because they did have some positive impact in our reaching a successful conclusion. We will thank them for their effort and stress we hope they gained some very valuable insights into the manufacturing business that will be helpful with other clients. Reminding them of the education they received from us is the only positive thing we can do. They certainly paid for it, but, it is hoped, it will pay off for them with some future client.

Now the fun part. We joyfully inform the winning vendor that they are the one we are willing to team with to build a fabulous future. We tell them why they won, and we subtly let them know we expect them to live up to our trust. We do all we can to get our message across that we are in it together and expect to work with them for quite a while. We will strive to build a winning team with them.

Step 9. It has been a long road, and we have finally arrived. We have an expert partner, with whom we feel comfortable, to help us build our particular CIM system. To get started, we map out an implementation plan based on our previously determined priorities. We make sure that our vendor can support those dates and activities, and, knowing as much as we do about their firm as they do about ours, we can be fairly confident that what we agree to can be accomplished.

Shoulder to shoulder we go forth to sell our plans to all members of our company. With the vendor's backstage assistance, we can now produce detailed and practical explanations of what benefits our CIM implementation will bring to our company.

Company X is now well on its way toward implementing CIM. The discovery and design phase has been successfully transited, and now the familiar project implementation task lies before them. The example I just presented is fairly typical of the thought processes and sequences companies go through while applying the prototype process. Simply, we can say that the prototype process is more involved than the ordinary way of purchasing software, but it ends up with a vastly superior product because it is precisely tailored to identified specific needs. I believe it is the only way to implement an optimized CIM system.

8. *Review vendor offerings of the integrated computer-based manufacturing system with the user community. Modify as required, and gain user acceptance.* During the prototype process, the systems council will do what can be referred to as a preliminary acceptability evaluation with selected company personnel. This is part of the vendor evaluation phase used to help the systems council select the right software and hardware supplier. At that stage the systems council only had their optimum flow and the vendors' proposals. Together these add up to proposed specifications. It is not an integrated computer-based manufacturing system (ICBMS). So, after a vendor is selected and has devised the ICBMS that the systems council accepts, the systems council tries it out by presenting it to the user community.

With the ICBMS we are far beyond the generalities of the optimum flow and have a very exacting, precisely documented set of software. The ICBMS is structured to meet all the demands of the optimum flow and the specifications. This is the next logical step after the prototype process, and the first iteration is usually very close to the system that will be implemented. It has been reviewed by the systems council and perhaps had to have some minor revisions made. The ICBMS to be presented to the users now has all the systems council's suggestions built into it, and from the council's point of view it is satisfactory.

The systems council has done about as much as it can to proof out the product the vendor is offering. Now they will unveil it to all the various users and get their feedback. Usually, this phase takes about a month of intensive presentations to the many user groups and requires detailed step-throughs of the relevant parts. Both the vendor and the systems council are after constructive criticism. They must know now if there are any serious glitches in the ICBMS as presently formulated. It is almost guaranteed that some modifications will have to be made, mostly minor and based on very subjective opinion. In other words, the vendor is just as correct as the user requesting the change. The important point is to make the change.

It is necessary for the users to "buy in." If the system is to work, the users have to have faith in it, or at least start out neutral. So, if relatively unimportant changes are requested, do them. It will solve many problems of confidence later on when the inevitable bugs surface. By conscientiously presenting the product to the people it was designed for, before it is installed, we take a giant step toward achieving acceptance. Remember, none of these users was part of the development process, yet they have to use the results. Therefore, they must be convinced that the product they are seeing, maybe for the first time, will actually make their job easier. Taking them into confidence, and making changes even though the work is virtually done, tends to prove to the users that the systems council and the vendor are not trying to force an autocratic system on them.

9. *Use good project management techniques to implement the integrated computer-based manufacturing system.* We are finally there. The CIM system is fully developed, evaluated, and tested as far as is practical. Now managers can do what they are comfortable doing, that is, implement a business decision. This step can take upward of a year. Recall that the prototype process usually requires less than a year. So, there should be no unbearable pressure to get it done, providing the plans are realistic and that sufficient implementation resources are made available. CIM normally has to be implemented within two to two-and-a-half years after the systems council has the authority to start phase 1. If it takes longer than that, senior management begins to lose interest and withdraws support. The process described in this chapter is compatible with that requirement.

The process of implementation involves virtually everything discussed in this book. In summary, it involves coordinating the software applications, training the users, installing hardware (mostly computer gear), and implementing the modules in actual use. This requires project management skills similar to any complex undertaking. Usually, CPM or PERT techniques are employed along with much communication to all levels within the organization, to keep people informed concerning status and benefits gained. I cannot overemphasize the need to communicate status of work. CIM is change, which is never readily accepted, and anything we can do to make it seem commonplace is beneficial. Communication does this and also builds up a fan following to cheer the team on to achieve the goal.

10. *Implement automated process equipment that is compatible with the integrated computer-based manufacturing system only after the basic system is operational.* We do not want "islands of automation." CIM is about optimizing the entire seven steps of the Manufacturing System, not just part of it. So, it is logical to put the framework in place first and then go about making enhancements that are truly beneficial.

A CIM system is driven by the communication requirements. If we have excellence in communication between the various functions of a business organization, then we are going to have a significantly better opportunity to maximize overall effectiveness than we would by independently implementing computer-driven process equipment. We do not want, nor can we tolerate, "islands of automation." We do want an optimized synergistic approach. For this reason, implementing CIM primarily emphasizes linkages of the seven steps of the Manufacturing System and only secondarily requires automated process equipment.

These ten phases represent what I call a strategy for implementing CIM. It is a commonsense approach of defining the present status and then using good management practices to investigate the feasibilities for improvement, selecting the most optimal pragmatic plan available, and going ahead to get the job done.

Implementing anything is, in essence, managing change. It takes careful planning and nurturing and the intestinal fortitude to say go and do it and then actually do it. I have no secret formulas to impart. I have only an engineer's experience of managing people and technology, and some advice to offer as a guide to implementing CIM. This book is full of advice on how to implement and use CIM. Let me close with one more learned truism. Nothing gets done without the good will of the people involved, no matter how noble the endeavor. Therefore, treat all those who will be touched by CIM with respect, dignity, and compassion, and it will all come out all right in the end.

Good luck and good hunting in the quest for making your company the shining star in the CIM universe.

Appendix A

Selected Related Readings

Aleksander, Igor. *Designing Intelligent Systems: An Introduction*. New York: Unipub, 1984.

Beeby, W., and Collier, P. *New Directions in CAD/CAM*. Dearborn, Mich.: Society of Manufacturing Engineers, 1986.

Blume, C., Dillmann, R., and Rembold, U. *Computer Integrated Manufacturing Technology and Systems*. New York: Marcel Dekker, 1985.

Chiantella, Nathan A., ed. *Management Guide for CIM*. Dearborn, Mich.: The Computer and Automated Systems Association of SME, 1986.

Childs, James J., *Principles of Numerical Control*. 3rd ed. New York: Industrial Press, 1982.

Chryssolouris, G., Francis P., and Von Turkovich, B., ed. *Manufacturing International '88: Symposium on Manufacturing Systems: Design, Integration, and Control*. New York: ASME Press, 1988.

Chryssolouris, G., and Jaikumar, R., ed. *Manufacturing International '88: Symposium on Management and Economics*. New York: ASME Press, 1988.

Chryssolouris, G., and Komanduri, R., ed. *Manufacturing International '88: Symposium on Product and Process Design*. New York: ASME, 1988.

Cox, J., and Goldratt, E. M. *The Goal: Excellence in Manufacturing*. Croton-On-Hudson, N.Y.: North River Press, 1984.

Dalziel, M. M., and Schoonover, S. C. *Changing Ways*. New York: Amacon, 1988.

Gessner, Robert A. *Manufacturing Information System, Implementation Planning*. New York: John Wiley, 1984.

Goddard, Walter E. *Just In Time: Surviving by Breaking Tradition*. Essex Junction, Vermont: Oliver Wight, 1986.

Gutowski, T. G., ed. *Manufacturing International '88: The Manufacturing Science of Composites*. New York: ASME Press, 1988.

Hall, Robert W. *Zero Inventories*. New York: Dow Jones, 1983.

Ham, I., Hitomi, K., and Yoshida, T. *Group Technology Applications to Production Management,* Boston, Mass.: Kluwer Nijhoff, 1985.

Harmon, P., and King, D. *Expert Systems*. New York: John Wiley, 1985.

Hartly, John. *FMS at Work*. New York: North-Holland, 1984.

Hyer, Nancy Lea, ed. *Capabilities of Group Technology*. Dearborn, Mich.: The Computer and Automated Systems Association of SME, 1987.

Jamshidi, M., Luh, J. Y. S., Seraji, H., and Starr, G. P., eds. *Robotics and Manufacturing: Recent Trends in Research, Education, and Applications*. New York: ASME Press, 1988.

Koenig, Daniel T. *Manufacturing Engineering: Principles for Optimization*. Washington, D.C.: Hemisphere, 1987.

Kops, L., ed. *Toward the Factory of the Future*. New York: ASME Press, 1980.

Koren, Yoram. *Computer Control of Manufacturing Systems*. New York: McGraw-Hill, 1983.

Lawrence, K. D., and Zanakis, S. H. *Production Planning and Scheduling*. Norcross, Ga.: Industrial Engineering and Management Press, 1984.

Lazarus, H., and Tomeshi, E. A. *People-Oriented Computer Systems: The Computer Crisis*. New York: Van Nostrand Reinhold, 1975.

Lublen, Richard T. *Just-in-Time Manufacturing: An Aggressive Manufacturing Strategy*. New York: McGraw-Hill, 1988.

Monden, Yasuhiro, ed. *Applying Just in Time: The American/Japanese Experience*. Norcross, Ga.: Industrial Engineering and Management Press, 1986.

Monden, Yasuhiro, *Toyota Production System*. Norcross, Ga.: Industrial Engineering and Management Press, 1983.

Muller, Thomas. *Automated Guided Vehicles*. New York: Springer-Verlag, 1983.

Pegden, C. D., and Pritsker, A. A. B. *Introduction to Simulation and Slam*. New York: John Wiley, 1979.

Prendergast, K. A., and Winston, P. H., ed. *The AI Business: Commercial Uses of Artificial Intelligence*. Cambridge, Mass.: MIT Press, 1984.

Riley, Frank J. *Assembly Automation: A Management Handbook*. New York: Industrial Press, 1983.

Shingo, Shigeo. *Non-Stock Production: The Shingo System for Continuous Improvement*. Cambridge, Mass.: Productivity Press, 1988.

Trappl, Robert. *Cybernetics: Theory and Applications*. Washington, D.C.: Hemisphere, 1983.

Tulkoff, J., ed. *CAPP: Computer-Aided Process Planning*. Dearborn, Mich.: Society of Manufacturing Engineers, 1985.

Wallace, Thomas F. *MRP II: Making Happen*. Essex Junction, Vermont: Oliver Wight, 1985.

Wight, Oliver W. *The Executives Guide to Successful MRP II*. Essex Junction, Vermont: Oliver Wight, 1982.

Wight, Oliver W. *Manufacturing Resource Planning: MRP II*. Essex Junction, Vermont: Oliver Wight, 1981.

Wight, Oliver W. *Production and Inventory Management in the Computer Age*. Boston, Mass.: CBI, 1974.

Index